RISK

UNCERTAINTY

and the

AGRICULTURAL FIRM

Charles B Moss
University of Florida, USA

RISK

UNCERTAINTY

and the

AGRICULTURAL FIRM

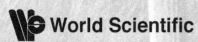

World Scientific

NEW JERSEY · LONDON · SINGAPORE · BEIJING · SHANGHAI · HONG KONG · TAIPEI · CHENNAI

Published by

World Scientific Publishing Co. Pte. Ltd.

5 Toh Tuck Link, Singapore 596224

USA office: 27 Warren Street, Suite 401-402, Hackensack, NJ 07601

UK office: 57 Shelton Street, Covent Garden, London WC2H 9HE

Library of Congress Cataloging-in-Publication Data
Moss, Charles B. (Charles Britt)
  Risk, uncertainty and the agricultural firm / by Charles B. Moss. -- 1st ed.
  p. cm.
  Includes bibliographical references.
  ISBN-13: 978-9814287623
  ISBN-10: 9814287628
  1. Farm management--Decision making--Mathematical models. 2. Risk--Mathematical models.
3. Uncertainty--Mathematical models. I. Title.

  S561.M68 2009
  630.68'4--dc22

                                                                    2009046295

British Library Cataloguing-in-Publication Data
A catalogue record for this book is available from the British Library.

Copyright © 2010 by World Scientific Publishing Co. Pte. Ltd.

All rights reserved. This book, or parts thereof, may not be reproduced in any form or by any means, electronic or mechanical, including photocopying, recording or any information storage and retrieval system now known or to be invented, without written permission from the Publisher.

For photocopying of material in this volume, please pay a copying fee through the Copyright Clearance Center, Inc., 222 Rosewood Drive, Danvers, MA 01923, USA. In this case permission to photocopy is not required from the publisher.

In-house Editor: Juliet Lee Ley Chin

Typeset by Stallion Press
Email: enquiries@stallionpress.com

Printed in Singapore.

Dedication

This book is dedicated to the memory of my grandparents Varney Eual Moss (1919–2008) and Opal Mae Moss (1917–2008) and the other Western Oklahoma sand land farmers who truly understood risk and the agricultural firm.

Risk and uncertainty is the basic reality of farming — Charles's home after a windstorm in the mid-1980s.

Preface

My family has a story about my grandmother, Opal Mae Moss. She and my grandfather, Varney Eual Moss, were sharecropping (or at least leasing under a share arrangement) from an old farmer by the name of Dudak, in Western Oklahoma (south of Sayre, Oklahoma) c. 1956. My grandmother wanted to buy a good dining room set, so she purchased about 100 turkey chicks to raise through the summer in order to earn money for the furniture purchase. The family story goes that in the fall, they lost the entire cotton crop in a hail storm. Grandma's turkey money had to be used to pay for as many of the cotton crop bills as possible, fund seed for the next year's crop, and provide their living for the year. Grandma did not get her dining room set until I lived on the farm in the early 1980s.

This story highlights several important aspects about the agricultural firm, but most importantly, it demonstrates that farming is inherently risky. Not only are prices and yields random functions of factors beyond a farmer's control, but also the farmer often has to make significant up-front investments (gambles) in order to farm for a living. My grandfather's apt saying "Cotton has paid more bills than any other crop in southwestern Oklahoma, but as long as you have cotton, you always have more bills to pay" is no exaggeration of the facts. In most cases, farming requires up-front money no matter what the yield or price becomes. The other interesting fact about the turkey story is that Grandma's turkey enterprise inadvertently became a risk management tool.

I left the farm for the academic life in the mid-1980s. The material presented in this book is the result of a class I teach at the University of Florida entitled "Agricultural Risk Analysis and Decision Making." The course serves both Master of Science and Ph.D. students from the Food & Resource Economics Department as well as other academic departments including Forestry and Animal Science. During the course presentation, I assume that the students have taken one mathematical microeconomics course and have a background in statistics or econometrics. The intent of the course is to provide a fairly general overview of the theory of decision making under risk with sufficient rigor to allow the students to approach the literature. In addition, I provide some of the nuts and bolts of risk analysis.

Any book project is a somewhat grueling undertaking, but I have been fortunate to have the support of numerous colleagues, mentors, and students. I would particularly like to recognize the Eminent Scholars at the University of Florida including the late Henri Theil (Economics), Andrew Schmitz (Food & Resource Economics), and Mark Flannery (Finance). I would like to thank several former graduate students for their diligent reading of earlier drafts of this manuscript, especially Grigorios Livanis and Matthew Salois. Finally, I would like to thank the two department chairs, Thomas Spreen and Ray Huffaker, under whose tenure this book was written.

Charles B. Moss

Contents

Preface   vii

1. Introduction   1

   1.1. Formulating the Risk Problem 3
   1.2. Decision Criteria 6
   1.3. Decision Making Under Risk:
       Fact and Fiction 8

2. Probability Theory — A Mathematical Basis for Making
   Decisions Under Risk and Uncertainty   10

   2.1. Set Theory and Probability 13
   2.2. Random Variables 15
   2.3. Conditional Probability and Independence 17
   2.4. Some Useful Distribution Functions 20
   2.5. Expected Value, Moments, and the Moment
       Generating Function 28
   2.6. Estimating Probability Functions 35
       2.6.1. Eliciting Subjective Probability
              Functions 36
       2.6.2. Objective Estimation of Probability
              Functions 37
       2.6.3. Production Functions and Trends 44
   2.7. Martingales and Random Walks 52
   2.8. Summary 56

3. Expected Utility — The Economic Basis of Decision
Making Under Risk   57

3.1. Consumption and Utility   58
3.2. Expected Utility   63
    3.2.1. Intuition behind Von Neumann and
           Morgenstern Proof   65
    3.2.2. Conceptual Structure of the Axiomatic
           Treatment of Numerical Utilities 67
    3.2.3. Analytical Examples of Expected
           Utility   69
3.3. Expected Value — Variance and Expected
     Utility Models   73
3.4. Problems with Expected Utility 76
3.5. Summary   82

4. Risk Aversion in the Large and Small   83

4.1. Arrow–Pratt Risk Aversion Coefficient   88
    4.1.1. Local Risk Aversion   88
    4.1.2. Transformations of Scale for the
           Arrow–Pratt Absolute Risk Aversion
           Coefficient   90
4.2. Eliciting Risk Aversion Coefficients 95
    4.2.1. Direct Elicitation of Utility Functions . . . 95
    4.2.2. Equally Likely Risky Prospect and Finding
           Its Certainty Equivalent (ELCE)   97
4.3. Summary   99

5. Portfolio Theory and Decision Making Under Risk   101

5.1. The Expected Value — Variance Frontier 106
5.2. A Simple Portfolio 108
5.3. A Graphical Depiction of the Expected
     Value–Variance Frontier   109
5.4. Mean–Variance versus Direct Utility
     Maximization   109

5.5. Derivation of the Expected Value–Variance
Frontier 112
    5.5.1. Derivation without a Risk-Free Asset . . . 112
    5.5.2. Derivation with a Risk-Free Asset 119
5.6. Summary 122

6. Whole Farm-Planning Models   124

6.1. Farm Portfolio Models 124
    6.1.1. Gains to Diversification Using Certainty
    Equivalence 128
    6.1.2. Extension to a Multiperiod Portfolio . . . 129
6.2. Minimize Total Absolute Deviation 131
6.3. Focus-Loss 136
6.4. Target MOTAD 138
6.5. Direct Utility Maximization 139
6.6. Discrete Sequential Stochastic Programming 141
6.7. Chance-Constrained Programming 143
6.8. Interpreting Shadow Values from Risk
Programming Models 146
6.9. Summary 148

7. Risk Efficiency Approaches — Stochastic Dominance   149

7.1. Stochastic Dominance 150
    7.1.1. The Concept of an Efficiency Criteria . . . 151
        7.1.1.1. First-Degree Stochastic
        Dominance 152
        7.1.1.2. Second-Degree Stochastic
        Dominance 152
    7.1.2. Increasing Risk 154
        7.1.2.1. Definition Based on Unanimous
        Preference 157
        7.1.2.2. Mean Preserving Spread 157
        7.1.2.3. Risk Aversion with Respect to a
        Function 162

7.2. Applications of Stochastic Dominance 166
7.3. Summary 172

8. Dynamic Decision Rules and the Value of Information 173

8.1. Decision Making and Bayesian Probabilities 173
8.2. Concepts of Information 177
8.3. A Model of Information 180
8.4. Summary 182

9. Market Models of Decision
   Making under Risk 183

9.1. Risk Equilibrium from the Consumer's Point
       of View 185
9.2. The Role of the Riskless Asset 189
9.3. Risk Equilibrium from the Firm's Perspective . . . 190
       9.3.1. Deriving the Security Market Line 191
       9.3.2. Supply of Stocks from the Firm 193
       9.3.3. Tests of the CAPM 194
       9.3.4. Incorporating Risk using CAPM 194
9.4. Arbitrage Pricing Theorem 196
       9.4.1. Single-Factor Model 198
       9.4.2. Two-Factor Model 200
9.5. Empirical Applications of Capital Market
       Models 202
       9.5.1. Capital Asset Pricing Models 202
       9.5.2. Tests for CAPM Efficiency 204
       9.5.3. Cross-Section Regression 205
       9.5.4. Arbitrage Pricing Model 206
9.6. Summary 208

10. Option Pricing Approaches to Risk 209

10.1. Introduction to Options and Futures 209
       10.1.1. Futures and the Hedge 209
       10.1.2. Options 211
       10.1.3. Option Pricing using Black–Scholes 215

10.2. Real Option Valuation   219

   10.2.1. Derivation of the Value of Waiting   220

   10.2.2. Application to Citrus   223

10.3. Crop Insurance   223

10.4. Summary   226

11. State Contingent Production Model: The Stochastic
Production Set   227

11.1. Depicting Risk and Input Decisions in the
Production Function   227

   11.1.1. Estimation of Stochastic Production
Functions using Quantile Regression . . .   228

   11.1.2. Developing the State-Space
Representation   234

11.2. State Production Set and Input
Requirement Set   237

11.3. Distance Functions and Risk Aversion   238

   11.3.1. Defining the Distance Function in
State-Contingency Space   238

   11.3.2. Risk Aversion and Valuing States   240

   11.3.3. Duality, Benefit, and Distance
Functions   241

   11.3.4. Defining Risk Aversion Graphically   243

   11.3.5. Constant Relative and Absolute Risk
Aversion   244

      11.3.5.1. Risk Premium   244

      11.3.5.2. Derivation of the Effort
Function   246

11.4. Summary   249

12. Risk, Uncertainty, and the Agricultural Firm — A
Summary and Outlook   250

Appendix A.   Measure Theory and the Justification
of Random Variables   253

Appendix B.   Derivation of the Moments of the Inverse
Hyperbolic Sine Distribution   257

Appendix C.   Numerical Techniques for Applied Optimization
and Solution of Nonlinear Systems of
Equations   265

Appendix D.   An Axiomatic Development of Expected Utility   268

Appendix E.   A GAMS Program to Select Optimal Portfolios   276

Appendix F.   R Program to Derive Optimum Portfolio with
and without a Risk-Free Asset   278

Appendix G.   Program to Compute the Efficient Frontier
with and without a Risk-Free Asset   281

Appendix H.   GAMS Program for the Portfolio Problem   283

Bibliography   285

Index   291

Chapter 1

Introduction

The study of economics by definition is the study of choices. Consumer demand is developed from the choices that consumers make based on their preferences and budget constraints. Supply and other production relationships are based on choices producers make in order to maximize their profit. Each set of decisions takes certain aspects of the decision problem as given. For example, in the consumption decision, the price of each good that can be consumed and the preference relationship (utility function) is taken as given. In the production problem, the price of the outputs and inputs and the production technology (or production function) are taken as given. Based on these assumptions, the consumer's demand functions, the output supply relationships and the input demand functions can be derived systematically using well-tested optimization theory. However, any student of economics will admit that the economy cannot be fully explained using this simplistic approach.

Complications arise from a variety of factors including institutional arrangements such as the assignment of property rights, government interventions through trade barriers and environmental restrictions, and incomplete or imperfect information. The fact that many decisions made by producers and consumers must be made with incomplete or imperfect information introduces a number of difficulties into the simple supply-and-demand framework. Specifically, the level of production of agricultural commodities is affected by variability of weather. Drought may cause corn yield in the midwestern United States to be lower than expected. The scarcity of corn will cause the price of corn to increase. The increased price of corn affects consumers in the United States in a variety of ways. The increased price of corn will be directly transferred to consumers by increases in the price of goods that are made directly from corn, such as corn chips. Corn is used as an input into the production process of other commodities that consumers may not associate with corn. For example, corn is a major input into the livestock industry. Thus, the increased price of corn increases the production cost in the beef, poultry, and dairy markets. The increased corn price would probably imply higher prices of meat, chicken, and milk. In the long run, if drought persists (such as in the dust-bowl of the 1930s), this will undoubtedly be true. However, in the short-run, the effect of increased corn prices may be more nebulous. For example, increased corn prices may cause beef producers to slaughter cattle earlier (rather than incurring the increased feed costs of feeding those cattle to a heavier weight). Similarly, increased corn prices may lead dairy producers to cull lower producing dairy cows from the herd, further decreasing the price of lower quality meats as the culled dairy cows are slaughtered. Finally, the increased price of corn may be partially offset by poultry producers as farmers shift from corn to less efficient, but cheaper alternative grains.

These price changes will filter through a host of secondary industries such as slaughter houses, wholesalers, and retailers each with differing abilities to react to the variation. For example, packing houses and wholesalers may have some ability to store beef. Hence, they may be able to draw on inventories to reduce the effect of a supply shortfall. It should be noted, however, that this moderation does not result from altruism, but a conscious activity to exploit a profit opportunity.

The litany of interactions described above is beyond the scope of this book, but it is presented to motivate an understanding of the complexity of the economic system. In such a complex system, certain facets of production and consumption are uncertain or at the very least unobservable. The simplest supply-and-demand diagrams proffered in introductory and intermediate microeconomics texts must be supplemented to develop a full understanding of the economic system. In this book, microeconomic foundations of risk analysis and its implications for decision making are emphasized. The topic development, thus, focuses on the effect of risk and uncertainty on firm decisions in general and agricultural firms in particular.

1.1. Formulating the Risk Problem

The first step in the development of any discussion is the establishment of terminology. In the case of decision making under risk, agreement on terminology is complicated by the variety of backgrounds and disciplines that have contributed to risk theory. As a starting point, consider the diagram of the decision process under uncertainty presented in Fig. 1.1.

The basis of the risk problem presented in Fig. 1.1 is the existence of two possible actions (a_1 and a_2). One example of such an action is for the farmer to apply either 90 pounds or 110 pounds of nitrogen per acre. If there is only one alternative, there is no decision problem (there are consequences of uncertainty, but not an economic problem[1]). The notion of alternative actions is not unique to decision

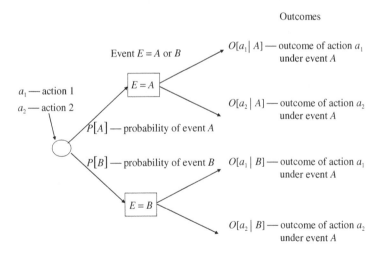

Fig. 1.1. Overview of the risk problem.

---

[1] This distinction is based on the definition that economics is the study of the allocation of scarce resources among unlimited and competing human wants and desires. If there is only one action, an allocation problem does not exist.

making under risk. In the deterministic production model, producers decide how much fertilizer to apply to an acre of corn. The difference is that the result of such a decision is known in advance or with certainty in the deterministic world. Thus, in a deterministic world the producer's problem of choosing the optimal level of nitrogen becomes

$$\max_{x_1} \pi = p_y x_1^\alpha x_2^{1-\alpha} - w_1 x_1 - w_2 x_2 \tag{1.1}$$

where π is the farmer profit, p_y is the price of the output (corn in this case), x_1 is the level of input 1 applied (nitrogen in this case), x_2 is the quantity of input 2 applied (other inputs that are held constant), w_1 is the price of the first input, and w_2 is the input price of the second input. The farmer's objective function or choice criteria in production economics is assumed to be to maximize profits. Thus, taking the first derivatives of Eq. (1.1) with respect to x_1 and solving for the level of x_1 that makes that derivative equal to zero yields an optimal action. Note there are two implicit assumptions in this process. First, it is assumed that the production function is deterministic (i.e., each level of input yields one and only one possible output). Second, it is assumed that the choice criterion was to maximize profit.

Diverging from the deterministic model, one may assume that the results (outcomes) of the farmer's actions are going to be affected by the state of nature of a random event (E). In Fig. 1.1, it is assumed that the event could result in two possible values $(A$ or $B)$. In this production scenario, these results will be qualitative (i.e., A is a high-rainfall event while B is a low-rainfall event). However, it is also possible to conceive of the outcome as a quantitative variable (i.e., E is the number of inches of rainfall received during the growing period). Also the event could be viewed as an input (albeit an uncontrollable input) into the production process. For example, let x_2 in Eq. (1.1) be the number of inches of rainfall received in the growing season. Note that the maximization problem in Eq. (1.1) assumes that the producer takes the level of x_2 as exogenous or fixed.

The next chapter develops the concept of probability more fully. For the purposes of the current discussion, assume that there is a number called a probability that characterizes the relative likelihood or frequency of an event (either $E = A$ or $E = B$ in Fig. 1.1) occurring. In the current example, assume that a high-rainfall outcome is relatively more likely than a low-rainfall event. Specifically, let the probability of a high-rainfall event equal to 0.60 (denoted $P[E = A] = 0.60$) and let the probability of a low-rainfall event equal to 0.40 (denoted $P[E = B] = 0.40$). Putting the entire framework together, assume that the value of the random event affects the outcome of the production decision based on the action taken.

Running through an empirical example, assume that the production of corn can be characterized by the Cobb–Douglas production function (as depicted in Eq. (1.1)), where x_1 is the choice of the level of nitrogen per acre and x_2 is the amount of rainfall received per year in inches. Given $\alpha = 0.30$, $p_y = 2.60$ and the price of nitrogen per pound at 0.05, the production results are presented in Table 1.1.

The combination of a random event and an action yields an outcome. Each outcome in Fig. 1.1 is denoted as $O[a_i|E = A]$ or the outcome of action a_i given that event A occurred. In Table 1.1, $O[a_1|E = A]$ denotes the outcome from applying 90 pounds of nitrogen per acre such that a high-rainfall event occurred (i.e., 35 in. of rain during the growing season). This outcome can be described in a variety of ways. $O[a_1|E = A]$ implies that the farmer produced 46.46 bushels of corn per acre. Alternatively, the outcome can be expressed in dollar space, or that this action resulted in a profit of $116.31/acre.

Table 1.1. Yield and profit per acre for the Cobb–Douglas production function.

Nitrogen per acre	Rainfall (inches per growing season)	
	30	35
	Corn yield (Bushels per acre)	
90	41.71	46.46
110	44.30	49.35
	Profit per acre ($s)	
90	103.95	116.31
110	109.68	122.80

1.2.   Decision Criteria

The forgoing discussion outlines the physical description of the economic problem facing the decision maker; however, at this point, it is devoid of economic content. This description in and of itself does not describe how the producer will choose between the two possible actions. Several conjectures will undoubtedly be offered by students familiar with economics. One that is usually offered is to choose the action that maximizes the expected value of production. Mathematically,

$$\max_{a_i} \begin{cases} E[\pi|a_1] = P[E = A]116.31 + P[E = B]103.95 = 111.37 \\ E[\pi|a_2] = P[E = A]122.80 + P[E = B]109.68 = 117.55 \end{cases}.$$

$$(1.2)$$

Under this decision criteria, a_2 is preferred to a_1 since $E[\pi|a_1] = 111.37 < E[\pi|a_2] = 117.55$. While this answer (i.e., maximizing the expected value of the production alternatives) would appear reasonable, students will probably be hard-pressed to give a compelling reason as to why such a criteria would be appropriate. One justification would be that since microeconomic theory assumes that consumers make decisions to maximize utility and since utility is constrained by income, then consumers would choose the action that makes them better off on average. This reasoning, however, ignores the broader question; namely, would consumers choose to be better-off on average, and if so, does maximizing the expected income make the consumer better-off on average? Looking forward to the topics in subsequent chapters, the expected utility hypothesis typically attributed to von Nuemann and Morgenstern (1953) demonstrates the conditions by which economic agents are conjectured to undertake actions that will maximize expected utility. However, only under the most stringent assumptions does maximizing expected income maximize expected utility.

A large portion of this book is dedicated to decision frameworks that either explicitly or implicitly assume that producers choose the action that either maximizes expected utility or produces an expected utility that dominates the expected utility generated by other alternatives. The latter group is typically referred to as risk efficiency criteria which eliminate inefficient, or inferior, alternatives. The expected utility formulation allows for the definition of several useful values. First, the expected utility of a risky alternative implies a certain amount of income that will leave the decision maker indifferent to the risky alternative. In other words, the certainty equivalent is the certain or fixed income that is equivalent to the risky outcome. Taken together, the certainty equivalent and the expected value of the risky alternative (or gamble) imply the maximum insurance premium that the decision maker is willing to pay to forgo the risk. While theoretically justified through the expected utility hypothesis, each of these amounts (the certainty equivalent and risk premium) depends on the decision maker's attitude about risk, typically referred to as the risk aversion. Individuals with higher risk aversion coefficients (to be discussed later) weight risk more heavily than individuals with lower risk aversion coefficients. More risk averse individuals have lower certainty equivalents (they are willing to take less in return for a risky alternative) and, thus, they are willing to pay a higher risk premium.

While certain procedures exist to elicit risk aversion coefficients, these coefficients are largely unobservable. As a result, several empirical approaches in risk analysis have been developed to operate under the assumption that individuals are risk averse, but the level of risk aversion itself cannot be observed. These procedures derive relationships that would exist between investment alternatives regardless of the exact level of risk aversion. Examples of this approach are frequently seen in financial economics. Both the Capital Asset Pricing Model (CAPM) and the Arbitrage Pricing Model (APM) begin with the assumption that investors are risk averse and then they derive relationships between asset prices based on arbitrage conditions (i.e., price relationships based on the ability to buy and sell stock in a way that leaves the risk of the asset portfolio unchanged). While these decision rules are robust with respect to risk aversion, their applicability in agriculture and other small firms are limited by the ability to arbitrage assets.

An alternative approach to risk analysis that does not depend on risk aversion is the option pricing approach. A financial option gives the holder the right to buy or sell an asset (a stock, bond, or commodity) at a specified price (strike price) at a specified date. Using the option pricing, approach analysts can derive the price of uncertainty using arbitrage arguments and assuming that decision makers are indifferent to risk (i.e., act in a way that maximizes the expected value of a risky alternative). Black and Scholes (1973) proposed a pricing rule that allows the analyst to empirically price a financial option based on arbitrage between the option contract, the cash market for the financial asset, and the futures market. Recent advances in the literature have reduced the need of actual arbitrage instruments, expanding the option valuation techniques beyond the boundaries of financial markets to include the valuation of other assets (Dixit and Pindyck, 1994).

1.3.   Decision Making Under Risk: Fact and Fiction

The forgoing discussion lays basic framework used to introduce decision making under risk and uncertainty. The discussion does not fit well into the classic normative versus positive paradigm of economic analysis. Many of the techniques have normative content in that they can be used to help decision makers make "better" decisions. However, many of the techniques are positive in that they can be used to predict firm level behavior in response to changes in agricultural policy, macroeconomic fluctuations, etc.

The normative dimension of risk analysis brings the most significant need for a caveat. In the words of a previous book on risk:

> Two potential misconceptions about our treatment should be immediately dismissed. First, a good risky decision does not guarantee a good outcome; rather, it is one consistent with the decision maker's beliefs about possible outcomes. A good decision is a considered choice based on a rational interpretation of the available information. Whether such a decision turns out right or wrong is partly a matter of luck and in any case it can never be determined until after the event, and often not even then. Second, the normative procedures we will present serve to aid decision makers through the complexities of their decision problems. The role of the decision maker is in

no way denigrated; as we shall see, his beliefs and preferences are of absolutely paramount importance.

(Anderson, Dillon, and Hardaker, 1977, p. 3)

From this discussion, it is important to emphasize that the techniques henceforth developed guarantee an appropriate or consistent decision, but do not guarantee a good result. Anecdotal evidence of errors that turned out well is the stuff of agricultural legend. For example, everyone has heard the story of a producer who held too much orange juice in inventory when a freeze occurred or of a producer who applied too much fertilizer in a year with higher than average rainfall. In each case, a poor decision had good consequences because of an anomaly. While such cases are the stuff of legend, it must be recognized that these decisions would have been disastrous in a typical year. Likewise, an appropriate level of fertilizer may yield a financial catastrophe in a severe drought. The best that can be hoped for is an appropriate, well-reasoned decision.

Second, the tools developed in this book do not replace the judgment of the decision maker, but augment them. Many of the tools require the decision maker's preferences (or risk aversion) and subjective judgment about the probability of outcomes. Looking forward to the next chapter, there are two different types of probabilities: objective and subjective. Objective probabilities are sometimes called frequencies. It is assumed that there is a physical process generating events that can be characterized by an objective (scientific) process. In this framework, the probability of a balanced coin landing heads up is 0.50. Subjective probabilities are based on the intuition and experience of the decision maker. For example, the decision maker may hold that the probability of a head is only 0.40 since the last result was a head. While the scientific community may hold the superiority of objective probability, it could be argued that subjective probabilities are superior since they are consistent with the individual's world view.

Chapter 2

Probability Theory — A Mathematical Basis for Making Decisions Under Risk and Uncertainty

The exposure to probability theory of most students comes in the guise of statistics. In statistics, researchers attempt to make an inference based on data that results from an experiment. In a drug trial, physicians may be interested in whether or not a drug had the intended effect. For example, did the new drug reduce the subject's cholesterol? Agronomists may conduct field experiments on crop yields: what is the effect of a genetic modification such as *Bacillus thuringiensis* (*Bt*) on corn yields under pest pressure? Each of these cases envision a classical hypothesis test (e.g., was this drug effective or did the introduction of a genetic modification improve corn yields?). The stuff of statistics is the formulation of procedures to determine whether any observed effect is significant or simply normal variations characteristic of the data. Such tests include discussions of confidence intervals, significance levels, etc. The applications in this text are not used to test hypotheses, but used to examine the effects of random variation on decision making. It is assumed in this discussion that planting decisions will have different results even when all controllable factors are held constant. Weather may affect corn yield (i.e., drought will reduce the corn yield). In addition, the crops may experience pests other than the one targeted by the genetic modification. Even though this text does not focus explicitly on statistical inference, some of the mechanics developed by statisticians can be used directly in risk analysis.

The use of statistical terms in risk analysis requires some degree of refinement in the general risk model presented in Chapter 1. Figure 1.1 uses two terms that could have conflicting meanings when used in risk analysis and statistical applications. Specifically, the term *event* could be used to denote an unknowable realization such as the quantity of rainfall during a growing season. The discussion then identifies the outcome as the result of an event and an action. In the case discussed in Chapter 1, the combination of a rainfall event and a fertilization decision was a crop yield outcome. In the vernacular of the statistician, the unknown or unknowable event is called a *random variable*. In this terminology, both the observed rainfall and the crop yield are outcomes of two different random processes or random variables. The observed value of a random variable is then referred to as an *observation* or *outcome* of the random variable. As will be discussed later in this chapter, a variety of relationships can be defined between two or more random variables. The simplest relationship is that the two outcomes are uncorrelated with each other. This relationship is referred to as *independence*. In the case discussed in Chapter 1, it is unlikely that the two events (rainfall observed in the growing season and corn yields) are independent. The physiology of plants implies that the yield is affected by the level of rainfall. From a statistical perspective, the choice of the level of fertilizer can be referred to as a *treatment*.

The overall mechanics is then that a random variable (or experiment) is defined as some event for which the outcome is unknown. If the outcome is known in advance then the process is not random, but *deterministic* or *certain*. An event whose outcome is not known with certainty is called *random* or *stochastic*. The set of all possible outcomes for a random variable is referred to as the *sample space*. Figure 1.1 presents one possible sample space. Specifically, the events depicted in Fig. 1.1 are *discrete*. A finite number of events (in this case two) are possible. Random variables with finite numbers of outcomes are typically referred to as *discrete random variables*. Discrete random variables are common in the discussion of statistics. For example, consider the outcome resulting from of tossing a fair coin. In this example, the sample space contains two elements (heads or tails). Likewise, the roll of a die yields six different outcomes; displaying one through six dots on each face.

The alternative to a discrete random variable is a *continuous random variable*. This random variable is best described by example. The discussion of crop yields in Chapter 1 considers only two possible outcomes for rainfall: high rainfall defined as 35 inches of rainfall in the growing season and low rainfall defined as 30 inches of rainfall in the growing season. Obviously, more than two levels of rainfall may occur in the growing season. Any positive amount of rainfall up to an unspecified upper-bound may occur. Logically, the possible set of rainfalls could be extended to include all the real numbers from zero bounded from above by a positive constant (b). Mathematically,

$$x \in [0, b) \qquad (2.1)$$

where x is a random variable denoting rainfall.

At first glance, it may be tempting to minimize the distinction between discrete and continuous random variables. However, the distinction is quite pronounced as the discussion is extended from random variables to probability. Intuitively, the *probability* of an event is defined as the relative likelihood that the event will occur. In the case of the coin toss, most people would conjecture that the probability of a coin toss coming up heads would be 0.50 (or $P[x =$ heads$] = 0.50$) where x is defined as the outcome of the coin toss. The question that then follows is: why is this proposition accepted? The standard reasoning is that two events of the discrete random variable are possible (heads or tails). If the coin is fair, it is anticipated that the two events are equally likely. Thus, the probability of heads would be

$$P[x = \text{heads}] = \frac{1\{x = \text{heads}\}}{1\{x = \text{heads}\} + 1\{x = \text{tails}\}} = \frac{1}{2} = 0.50. \quad (2.2)$$

This interpretation is subjective in that the probabilities are assumed based on expectations without data, but the intuition follows the *frequency approach* to probability or the assignment of probability based on the frequency that an event is observed. Technically, the relative frequency of an event is defined as

$$P[x] = \frac{f}{N} \tag{2.3}$$

where f is the frequency or number of times that an outcome occurs and N is the total number of times an event is observed. An objective approach would then be to use the relative frequency of observed events to define the probability of a discrete random variable.

The nuance is introduced by the fact that no event is likely to repeat itself if the random variable is continuous. For example, one rainfall event could be 30 inches while the rainfall the next year could be 30.1 inches. In the limit (i.e., over a large number of observations), the frequency approach would yield a zero level of probability for each individual event.

2.1.   Set Theory and Probability

The foundations of probability are based, like most branches of mathematics, on set theory. While a complete discussion of this foundation is beyond the scope of this text, some basic notions are useful in the development of decision making under risk.

Most are familiar with the rudiments of set theory. A *set* is defined as a collection of objects. In the current discussion, the set of all possible outcomes for the coin toss includes two objects (heads or tails). In reverse, an *element* is an object that belongs to the set. To be slightly more rigorous, let $x = 1$ if the outcome of a coin toss is a head and $x = 0$ if the outcome is a tail. Thus,

$$C \equiv \{x = 1, 0\} \Leftrightarrow x = 1 \in C \quad \text{and} \quad x = 0 \in C. \tag{2.4}$$

More complicated examples of elements include the set of positive real numbers. In our discussion of corn yields, $x = 41.71 \in R^+$ where R^+ denotes the set of all positive real numbers.

Definition 2.1. A set that contains no elements is referred to as a *null set* and denoted $C = \varnothing$ or $C = \{ \ \}$.

Definition 2.2. The *union* of two sets C_1 and C_2 is the set that contains any element that is contained in either set or both: $x \in C \equiv C_1 \cup C_2$ if $x \in C_1$, $x \in C_2$, or $x \in C_1$ and $x \in C_2$.

Definition 2.3. The *intersection* of two sets is that set that contains any element that is contained in both sets: $C_1 \cap C_2 \equiv \{x : x \in C_1$ and $x \in C_2\}$.

Both unions and intersections can be extended to more than two sets:

$$\bigcup_{i=1}^{n} C_i = C_1 \cup C_2 \cdots C_n$$

$$\bigcap_{i=1}^{n} C_i = C_1 \cup C_2 \cdots C_n. \tag{2.5}$$

In order to extend the discussion, the concept of containment is introduced. First, a group of sets are defined over the same space. For example, let $C_1, C_2, \ldots$ be defined as subsets within a larger set or space C.

Definition 2.4. A *subset* C_i $(i = 1, \ldots, n)$ is defined on a larger set C if every element that belongs to C_i is also a member of $C : C_i \subset C$ if $x \in C_i$ implies $x \in C$.

Given this definition of containment the complement of the set can then be defined:

Definition 2.5. The *complement* of C_i (denoted C_i^c) is defined as those elements of C that are not elements of C_i: $C_i^c \equiv \{x : x \in C$ but $x \notin C_i\}$.

The *probability function* is then defined as a measure on a space of possible events (Appendix A provides an introduction to measures defined on sets). This measure is dependent on the mechanics related to fields. Thus, the next step in the process is to define a *σ-field*.

Definition 2.6. If B is a collection of subsets of C, B is a *σ*-field given

(1) $\varnothing \in B$ (thus B is not empty),

(2) If $C_i \in B$ then $C_i^c \in B$ (B is closed under complementation), and

(3) If the sequence of sets $\{C_1, C_2, \ldots\}$ is in B then $\bigcup_{i=1}^n C_i \in B$ (B is closed under countable union).

The definition of a probability measure is based on three axioms. To motivate the intuition behind these properties, consider the intuitive properties of relative frequencies. Returning to Eq. (2.3)

$$P[x] = f_C(x) = \frac{f}{N} \tag{2.6}$$

where $f_C(x)$ is defined as the relative frequency of event $x \in C$ after N repetitions. Note that $0 \le f_C(x) \le 1$. First, since $f \ge 0$ it is obvious by Eq. (2.6) that $f_C(x) \ge 0$. Second, it should be obvious that $f \le N$ (i.e., more heads cannot be observed than coin tosses) so that $f_C(x) \le 1$. Thus, the definition of a probability measure follows:

Definition 2.7. If C is the sample space and B is a σ-field on C, then $P[x]$ is defined as a real valued function on B. $P[x]$ is a *probability set function* that satisfies three conditions:

(1) $P[x] \ge 0$ for all $x \in B$,

(2) $P[C] = 1$, and

(3) If $\{C_i\}$ is a sequence of sets in B and $C_n \cap C_m = \varnothing$ for all $m \ne n$ then

$$P\left[\bigcup_{i=1}^{\infty} C_i \right] = \sum_{i=1}^{\infty} P[C_i].$$

The last part of the definition gives rise to the fact that

$$P[C_i] = 1 - P[C_i^c] \tag{2.7}$$

since $C_i \cap C_i^c = \varnothing$ by definition of the complement and $C_i \cup C_i^c = C$.

2.2. Random Variables

Given the definition of probability, a formal definition of random variables can be developed. At one level, the random variable is obvious; it is the number of dots on the top of the die. Stepping back a little, consider the outcome of planting a crop. In production economics, a production function is used to model the result of planting a crop. In its most general manifestation, the production function can be depicted as a mapping function between a set of inputs and an output

$$F : R_n^+ \rightarrow R_1^+ \tag{2.8}$$

where F is the production function, R_n^+ denotes the use of a non-negative set of n inputs, and R_1^+ is a non-negative single output. One example of this function is the Cobb–Douglas production function implicit in Eq. (1.1). Using a slightly more general form of this function

$$y = F(x) = x_1^\alpha x_2^\beta x_3^\gamma \Rightarrow F : R_3^+ \rightarrow R_1^+ \tag{2.9}$$

where x_1, x_2, and x_3 are inputs and α, β, and γ are parameters.

Given the functional mapping framework, a random variable can be defined as

$$X : C \rightarrow R_1^+ \tag{2.10}$$

where C is the sample space from Definition 2.7. This event space could be simple or complex (i.e., a simple event is an event that cannot be decomposed into smaller parts such as the value of the roll of a single die and a complex event is anything that is not a simple event). As an example, C could be a vector of weather variables such as heat and moisture that contribute to an observed pest pressure which defines the random variable (e.g., X could be the number of grasshoppers). The probability function or measure is then defined based on that random variable

$$P_X[X = x_i] = P(\{C_i \in C, X(C_i) = x_i\}). \tag{2.11}$$

where $P_X[\ ]$ denotes the probability defined on the output of the random variable [implicitly $X(C_i) = x_i \Rightarrow C_i = X^{-1}(x_i)$]. Two alternative definitions of a random variable follow

Definition 2.8. A *random variable* is a variable that takes values according to a certain probability.

Definition 2.9. A *random variable* is a real-valued function defined over a sample space.

Following the axioms of probability:

(1) $P[X = x_i] \geq 0$ for all x_i,
(2) If $x_1, x_2, \ldots$ are mutually exclusive outcomes of the random process
     $P[\bigcup_{i=1}^{n} x_i] = \sum_{i=1}^{n} P_X[X = x_i]$, and
(3) $\sum_i P_X[X = x_i] = P[C] = 1$.

Thus, the definition of random variables follows the axioms of probability.

Moving closer to a pragmatic application, the distribution function of a continuous random variable is defined as

$$P_X[X = x] = f(x) \tag{2.12}$$

where $f(x) \geq 0$ for all $x \in X$ or the range of possible values of the random variable and

$$\int_{x \in X} f(x)dx = 1. \tag{2.13}$$

2.3.   Conditional Probability and Independence

In order to define the concept of a *conditional probability*, it is necessary to define *joint* and *marginal probabilities*. Further, the simplest way to define joint and marginal probabilities is by example, usually using games of chance. To this end consider rolling two die. The outcome of both die can be represented as a joint event. For example, the outcome of 4 on the first die and 6 on the second die is a joint event $\{x_1, x_2\} = \{4, 6\}$. Intuitively, the probabilities of this event occurring is $1/36$. Thus, the joint probability is then the probability of an ordered pair. A marginal probability is the probability of one outcome regardless of the outcome of the other event. For example, what is the probability of the first die being 4 or $P[x_1 = 4]$? This is the sum of six different probabilities of the joint event (the joint events being $\{4, 1\}$, $\{4, 2\}$, $\{4, 3\}$, $\{4, 4\}$, $\{4, 5\}$, and $\{4, 6\}$). Mathematically assuming that the outcome of each joint event is equally likely yields

$$P[x_1 = 4] = P[\{4, 1\}] + P[\{4, 2\}] + P[\{4, 3\}]$$
$$+ P[\{4, 4\}] + P[\{4, 5\}] + P[\{4, 6\}]$$
$$= \frac{1}{36} + \frac{1}{36} + \frac{1}{36} + \frac{1}{36} + \frac{1}{36} + \frac{1}{36}$$
$$= \frac{6}{36} = \frac{1}{6}. \tag{2.14}$$

The conditional probability is then the probability of one event, such as the probability that the first die is a 4, given that the value of another random variable is known, such as the fact that the value of the second die roll is equal to 6. In the forgoing example, the case of the fair die, this value is $1/6$ (or $P[x_1 = 4 | x_2 = 6] = 1/6$).

Following the axiomatic approach to defining probability, the axioms of conditional probability can be defined as:

Definition 2.10. Given that A and B are sets defined on C, the Axioms of Conditional Probability are

(1) $P[A|B] \geq 0$ for all A,
(2) $P[A|A] = 1$,
(3) If $\{A_i \cap B_i\}i = 1, 2, \ldots$ are mutually exclusive events, then $P[A_1 \cup A_2 \cup \cdots \cup A_n | B] = P[A_1|B] + P[A_2|B] + \cdots + P[A_n|B]$, and
(4) If $B \supset H$, $B \supset G$, and $P[G] \neq 0$ then $\frac{P[H|B]}{P[G|B]} = \frac{P[H]}{P[G]}$.

The first three conditions follow the general axioms of probability theory. The final condition states that the relative probability of a conditional event and marginal distribution functions are the same. Given this construction, the conditional probability can be derived as

$$P[A|B] = \frac{P[A \cap B]}{P[B]}. \tag{2.15}$$

In the discussion of the role of the die

$$P[x_1 = 4 | x_2 = 6] = \frac{P[x_1 = 4 \text{ and } x_2 = 6]}{P[x_2 = 6]} = \frac{\frac{1}{36}}{\frac{1}{6}} = \frac{1}{6}. \tag{2.16}$$

Given that the events $A_1, A_2, \ldots, A_n$ are mutually exclusive events such that $P[A_1 \cup A_2 \cup \cdots \cup A_n] = 1$ the conditional probability can then be extended to *Bayes Theorem*

$$P[A_i|E] = \frac{P[E|A_i]P[A_i]}{\sum_{j=1}^{n} P[E|A_j]P[A_j]} \qquad (2.17)$$

where E is a conditioning event. Note, turning Eq. (2.15) around

$$P[A|B] = \frac{P[A \cap B]}{P[B]} \Rightarrow P[B]P[A|B] = P[A \cap B]. \qquad (2.18)$$

Substituting this result into Eq. (2.17)

$$P[A_i|E] = \frac{P[A_i \cap E]}{\sum_{j=1}^{n} P[A_j \cap E]}. \qquad (2.19)$$

This expression unifies the simple expression in Eq. (2.15). Specifically, following the Axioms of Probability (Condition 3)

$$P[A_1] + P[A_2] + \cdots + P[A_n] = P[A_1 \cup A_2 \cup \cdots \cup A_n]. \qquad (2.20)$$

Thus, the denominator of Eq. (2.19) becomes

$$P[A_1 \cap E] + P[A_2 \cap E] + \cdots + P[A_n \cap E]$$
$$= P[(A_1 \cup A_2 \cup \cdots \cup A_n) \cap E]. \qquad (2.21)$$

Given that $A_1, A_2, \ldots, A_n$ are exhaustive, or $A_1 \cup A_2 \cup \cdots \cup A_n = C$ since $P[A_1 \cup A_2 \cup \cdots \cup A_n] = 1$

$$P[(A_1 \cup A_2 \cup \cdots \cup A_n) \cap E] = P[C \cap E] = P[E] \qquad (2.22)$$

with the last equality since $E \subset C$. Returning to the numerator in Eq. (2.17)

$$P(E) = \sum_{i=1}^{n} P[E|A_i]P[A_i]. \qquad (2.23)$$

Given this superstructure the relationship between sets of random variables can be further defined. Specifically, the definition of conditional distribution allows for the definition of statistical independence.

Definition 2.11. Two events A and B are independent if $P[A] = P[A|B]$.

Thus, in our discussion of rolling two die above

$$P[x_1 = 4] = P[x_1 = 4|x_2 = 6] = \frac{1}{6}. \qquad (2.24)$$

Three events A, B, and C are said to be independent if

(1) $P[A \cap B] = P[A]P[B]$,
(2) $P[A \cap C] = P[A]P[C]$,
(3) $P[B \cap C] = P[B]P[C]$, and
(4) $P[A \cap B \cap C] = P[A]P[B]P[C]$.

2.4. Some Useful Distribution Functions

The array of distribution functions available for use in decision making under risk is bewildering, but a smaller set of well-tested distributions are typically used as a basis for empirical work. Among these distributions, the *normal distribution* stands out for several reasons. First, the use of the normal distribution can sometimes be justified on statistical grounds alone since the central limit theorem indicates that many variables of interest are asymptotically normal. Second, the multivariate form of the normal distribution allows for easily quantifiable dependence or independence of multivariate distributions. Third, the parameters of this distribution are easily estimated from empirical data. The univariate form of the normal distribution is written as

$$f(x|\mu, \sigma^2) = \frac{1}{\sigma\sqrt{2\pi}} \exp\left[-\frac{(x - \mu)^2}{2\sigma^2}\right] \qquad (2.25)$$

where x is the random variable, σ^2 is the variance of the distribution, and μ is the mean of the distribution. The multivariate form of the normal distribution is written as

$$f(x|\mu, \Sigma) = \frac{1}{\sqrt{2\pi}} |\Sigma|^{-\frac{1}{2}} \exp\left(-\frac{1}{2}(x - \mu)'\Sigma^{-1}(x - \mu)\right) \qquad (2.26)$$

where x is a vector of outcomes ($x = (x_1 \quad \cdots \quad x_n)'$) and Σ is a positive definite (or possibly positive semi-definite) variance matrix

$$\Sigma = \begin{bmatrix} \sigma_{11} & \sigma_{12} & \cdots & \sigma_{1n} \\ \sigma_{21} & \sigma_{22} & \cdots & \sigma_{2n} \\ \vdots & \vdots & \ddots & \vdots \\ \sigma_{n1} & \sigma_{n2} & \cdots & \sigma_{nn} \end{bmatrix}. \tag{2.27}$$

The diagonal elements of the variance matrix ($\sigma_{ii}, i = 1, \ldots, n$) in the multivariate normal distribution are the variances from the univariate distributions in Eq. (2.25). The off-diagonal elements ($\sigma_{ij}, i \neq j$) are the *covariances* which are related to the correlation coefficients and they contain information about the independence between the two random variables. Specifically, the *correlation coefficient* (ρ_{ij}) can be defined as

$$\rho_{ij} = \frac{\sigma_{ij}}{\sqrt{\sigma_{ii}\sigma_{jj}}}. \tag{2.28}$$

If normal random variables are independent, they are uncorrelated, and their covariance is zero.

Another commonly used continuous distribution function is the *uniform distribution*. The advantages of this distribution include its simplicity. Specifically, the most general form of the uniform distribution is

$$f(x|a, b) = \begin{cases} \dfrac{1}{b-a} & \text{if } x \in [a, b] \\ 0 & \text{otherwise} \end{cases} \tag{2.29}$$

where a and b are known constants. To demonstrate the simplicity of the uniform distribution, the conjecture in Eq. (2.13) will be proven (i.e., that the total probability of the sample space equals 1)

$$\int_{-\infty}^{\infty} f(x|a, b)dx = \int_{-\infty}^{a} 0\, dx + \int_{a}^{b} \frac{1}{b-a}dx + \int_{b}^{\infty} 0\, dx$$

$$= \left(\frac{x}{b-a} \Big|_{a}^{b} \right) = \frac{b}{b-a} - \frac{a}{b-a} = \frac{b-a}{b-a} = 1. \tag{2.30}$$

In addition, the uniform distribution can be used to demonstrate the independence properties. Taking $a = 0$ and $b = 1$, the bivariate form of the uniform distribution becomes

$$f(x_1, x_2) = \begin{cases} 1 & \text{if } 0 \leq x_1 \leq 1 \text{ and } 0 \leq x_2 \leq 1 \\ 0 & \text{otherwise} \end{cases}. \tag{2.31}$$

Note that the multivariate distribution obeys the result depicted in Eq. (2.13) in that

$$\int_0^1 \int_0^1 f(x_1, x_2) dx_1\, dx_2 = \int_0^1 ((x|_0^1) dx_2 = \int_0^1 (1-0) dx_2$$
$$= \int_0^1 dx_2 = (x|_0^1 = 1 - 0 = 1. \tag{2.32}$$

To examine the conditional properties of the bivariate uniform distribution, the *marginal distribution* of x_1 is derived. This marginal distribution is derived by integrating out x_2 (i.e., integrating across all possible values of x_2)

$$f_1(x_1) = \int_0^1 f(x_1, x_2) dx_2 = \int_0^1 dx_2 = (x_2|_0^1 = 1. \tag{2.33}$$

Note also that $f_2(x_2) = 1$. The *conditional distribution* for x_2 given the value of x_1 is then written as

$$f(x_2|x_1) = \frac{f(x_1, x_2)}{f_1(x_1)} = \frac{1}{1} = 1. \tag{2.34}$$

Therefore x_1 and x_2 are independent since

$$f_2(x_2) = f(x_2|x_1) = 1. \tag{2.35}$$

A third distribution function sometimes used in risk and decision making for the agricultural firm is the *gamma distribution* (Collender and

Zilberman, 1985)

$$f(x|\alpha,\beta) = \begin{cases} \dfrac{1}{\Gamma(\alpha)\beta^\alpha}x^{\alpha-1}\exp(-x/\beta) & \text{for } 0 < x < \infty \text{ and } \beta > 0 \\ 0 & \text{otherwise} \end{cases}$$

$$(2.36)$$

where $\Gamma(\alpha)$ is the gamma function. This distribution function is preferred to the normal distribution in the characterization of crop yield distributions because its range includes only positive values. The shape of the distribution imposes very specific characteristics on the crop distribution, which will be discussed later.

Another distribution that has been used to model the randomness of crop production is the *beta distribution* (Nelson and Preckel, 1989)

$$f(x|\alpha,\beta) = \begin{cases} \dfrac{1}{\mathrm{B}(\alpha,\beta)}x^{\alpha-1}(1-x)^{\beta-1} & \text{for } 0 < x < 1, \ \alpha > 0, \ \beta > 0 \\ 0 & \text{otherwise} \end{cases}$$

$$(2.37)$$

where $\mathrm{B}(\alpha,\beta)$ is the beta function. One difficulty with using the beta function to model crop yields is the fact that the random variable (x) lies between 0 and 1. Thus, the sample must be scaled by dividing by an estimate of the maximum crop yield.

One alternative to using these standard distributions involves the use of *transformation functions* to create a new distribution. As a starting point, remember the basic property of the distribution function from Eq. (2.13). Specifically, the cumulative distribution of a random variable can be written as

$$F(x^*) = \int_{-\infty}^{x^*} f(x)dx \Leftrightarrow P[x \leq x^*] \qquad (2.38)$$

where x^* is some fixed value of the random variable. Next, consider a transformation function such that

$$z = \phi(x) \qquad (2.39)$$

so that the value of the random variable results from a transformation of another random variable. Given that $\phi(x)$ is a one-to-one mapping from x (so that $\phi^{-1}(z)$ is well defined) into z, substitute this transformation directly into the cumulative density function $F[\phi^{-1}(z)]$. Differentiating the cumulative density function yields the probability density function for the transformation

$$g(z) = f(\phi^{-1}(z)) \left| \frac{d\phi^{-1}(z)}{dz} \right| \tag{2.40}$$

where $g(z)$ is the probability density function of the variable that is transformed into the normal distribution. The absolute value of the $d\phi^{-1}(z)/dz$ is used because the one-to-one transformation function could be either increasing (so that $d\phi^{-1}(z)/dz \geq 0$) or decreasing (so that $d\phi^{-1}(z)/dz \leq 0$). In either case, since $f(\ )$ is always positive by the definition of a probability density function, the absolute value transformation guarantees that the resulting probability density function is positive throughout its range.

There are numerous applications to the transformation approach. One application is its use to model correlated non-normal random variables. Moss and Shonkwiler (1993) and Ramirez, Moss, and Boggess (1994) use the inverse hyperbolic sine to transform a non-normal random variable to normality. Specifically, both studies use the transformation

$$e_t = e_t(\varepsilon_t, \theta) = \frac{\ln(\theta\varepsilon_t + [(\theta\varepsilon_t)^2 + 1]^{\frac{1}{2}})}{\theta} \tag{2.41}$$

where e_t is a normally distributed random variable, ε_t is the non-normal random variable, and θ is a parameter of the transformation. Following this definition, the distribution of ε_t becomes

$$g(\varepsilon_t | \mu, \sigma^2, \theta) = \frac{1}{\sigma\sqrt{2\pi}} \exp\left[-\frac{1}{2} \frac{(e_t(\varepsilon_t, \theta) - \mu)^2}{\sigma^2} \right] (e_t^2(\varepsilon_t, \theta)\theta^2 + 1)^{-\frac{1}{2}}. \tag{2.42}$$

Ramirez, Moss, and Boggess expand the formulation to a vector of random variables

$$e_{1,t}(\varepsilon_{1,t}, \theta_1) = \frac{\ln(\theta_1\varepsilon_{1,t} + [(\theta_1\varepsilon_{1,t})^2 + 1]^{\frac{1}{2}})}{\theta_1}$$

$$e_{2,t}(\varepsilon_{2,t}, \theta_2) = \frac{\ln(\theta_2\varepsilon_{2,t} + [(\theta_2\varepsilon_{2,t})^2 + 1]^{\frac{1}{2}})}{\theta_2} \tag{2.43}$$

so that plugging the results of Eq. (2.43) into the definition of the multivariate normal in Eq. (2.26) and multiplying by the Jacobian terms yield the distribution function for the vector ε_t

$$g(\varepsilon_t|\mu, \Sigma, \theta) = \frac{1}{\sqrt{2\pi}}|\Sigma|^{-\frac{1}{2}} \exp\left[-\frac{1}{2}(e_t(\varepsilon_t, \theta) - \mu)'\Sigma^{-1}(e_t(\varepsilon_t, \theta) - \mu)\right]$$

$$\times (e_{1,t}^2(\varepsilon_{1,t}, \theta_1)\theta_1^2 + 1)^{-\frac{1}{2}}(e_{2,t}^2(\varepsilon_{2,t}, \theta_2)\theta_2^2 + 1)^{-\frac{1}{2}}.$$

$$(2.44)$$

Appendix B presents a detailed discussion of the inverse hyperbolic sine transformation.

Looking forward to a discussion of subjective probability, the *triangular probability density function* is one alternative for eliciting a subjective probability density function from producers. The triangular probability density function only requires the decision maker to specify the minimum (x_1), *mode* (or most likely x_M), and maximum outcome (x_2). The probability density function is then drawn as depicted in Fig. 2.1.

Algebraically, the triangular probability density function is a spline function, which can be written as

$$f(x) = \begin{cases} 0 & \text{if } x < x_1 \\ a + bx & \text{if } x_1 \le x < x_M \\ c + dx & \text{if } x_M \le x < x_2 \\ 0 & \text{if } x \ge x_M \end{cases} \qquad (2.45)$$

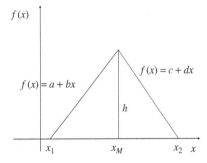

Fig. 2.1. Triangular probability density function.

where a, b, c, and d are constants. To solve for these constants, note that the height of each segment of the probability density function (h) is equal at the mode. In addition, by the definition of the probability function, the total area under the triangular probability density function is 1 (i.e., Eq. (2.13)). In this case, the area under the triangular probability density function is the sum of the area of two triangles depicted in Fig. 2.1 (i.e., the first triangle is $1/2(x_M - x_1)h$ while the area of the second triangle is $1/2(x_2 - x_M)h$). Thus, solving the height of the probability density function at the mode yields

$$\frac{1}{2(x_M - x_1)h} + \frac{1}{2(x_2 - x_M)h} = 1$$

$$h = \frac{2}{(x_m - x_1) + (x_2 - x_M)} = \frac{2}{x_2 - x_1}. \tag{2.46}$$

Solving for the slope in the first spline yields

$$b = \frac{\frac{2}{(x_2 - x_1)}}{(x_M - x_1)} = \frac{2}{(x_M - x_1)(x_2 - x_1)} \tag{2.47}$$

while solving for the constant in that spline gives

$$f(x_1) = a + bx_1 = 0 \Rightarrow a = -bx_1. \tag{2.48}$$

Next, solving for the slope of the second spline

$$d = -\frac{\frac{2}{(x_2 - x_1)}}{(x_2 - x_M)} = -\frac{2}{(x_2 - x_M)(x_2 - x_1)}. \tag{2.49}$$

Solving for the constant in the second spline

$$f(x_2) = c + dx_2 = 0 \Rightarrow c = -dx_2. \tag{2.50}$$

As an empirical example, consider the problem of eliciting a probability density function for the price of corn in the next harvest period. First, ask the producer what he anticipates the lowest possible corn price will be and he responds \$1.75/bushel. Next, ask what he believes the highest possible corn price could be at the next harvest and he responds \$3.25/bushel. Finally, ask what the most likely corn price during the next harvest period is and he responds $2.30/bushel. Based on these three answers, calculate

$$b = \frac{2}{3.25 - 1.75} = \frac{4}{3}. \tag{2.51}$$

Next, calculate the slope of the first spline

$$b = \frac{2}{(2.30 - 1.75)(3.25 - 1.75)} = 2.42424. \tag{2.52}$$

Completing the first spline

$$a = -\frac{4}{3}(2.4242) = -4.24242. \tag{2.53}$$

Solving for the slope of the second spline

$$d = -\frac{2}{(3.25 - 2.30)(3.25 - 1.75)} = -1.40351. \tag{2.54}$$

Solving for the constant of the second spline

$$c = -1.40351(3.25) = 4.5614. \tag{2.55}$$

This distribution is depicted in Fig. 2.2.

Apart from the ability to solicit probability density functions from producers, the triangular and modifications of the triangular distribution have been used in the literature. Featherstone *et al.* (1988) use

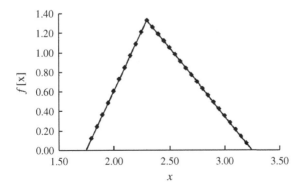

Fig. 2.2. Empirical example of triangular probability density function.

the triangular distribution function and a left-truncated triangular distribution function to model the effect of price floors on optimal debt at the farm level.

2.5.   Expected Value, Moments, and the Moment Generating Function

The forgoing discussion presents one statistical description of random variables. However, while the probability density function provides a complete description of the random variable, the random variable can also be described using a statistic. Technically, a *statistic* is result or formula that can be used to summarize information about the probability density function. One basic statistic encountered by students in statistics courses is the mean of a random variable (also referred to as the expected value of the distribution). From a mathematical point of view, there are two ways to develop a mean. The first definition of a mean results from distribution or probability density function for the random variable described in the preceding section.

Definition 2.12. The *expected value* (*expectation* or *mean*) of a discrete random variable X, denoted $E[X]$, is defined as $E[X] = \sum_{x_i \in X} x_i P[x_i]$ if the sequence $\sum_{x_i \in X} x_i P[x_i] < \infty$ (or that the measure is bounded).

Definition 2.13. The *expected value* of a continuous random variable X is then defined as $E[X] = \int_{x \in X} x f(x) dx$ if the sequence $\int_{x \in X} x f(x) dx < \infty$ (or the measure is bounded).

Taking the die roll as an example, if we let each side be equally likely the expected value of the roll of a fair die is

$$E[x] = \sum_{i=1}^{6} i P[i] = 1\frac{1}{6} + 2\frac{1}{6} + 3\frac{1}{6} + 4\frac{1}{6} + 5\frac{1}{6} + 6\frac{1}{6} = \frac{21}{6} = 3\frac{1}{2}.$$

$$(2.56)$$

This result points out an interesting fact about the expected value, namely that the expected value need not be an element of the sample set. Next, suppose we weight the die so that it is no longer fair. Specifically, assume that $P[i] = 1/9$ for $i = \{1, 2, 5, 6\}$, $P[3] = 3/9\{= 1/3\}$, and $P[4] = 2/9$, then the expected value becomes

$$E[X] = \sum_{i=1}^{6} iP[i] = 1\frac{1}{9} + 2\frac{1}{9} + 3\frac{3}{9} + 4\frac{2}{9} + 5\frac{1}{9} + 6\frac{1}{9} = \frac{31}{9}.$$

$$(2.57)$$

Taking the uniform distribution as an example, the mean is derived as

$$E[X] = \int_{0}^{1} xf(x)dx = \int_{0}^{1} x\, dx = \left(\frac{1}{2}x^2\Big|_{0}^{1}\right) = \frac{1}{2}(1 - 0) = \frac{1}{2}.$$

$$(2.58)$$

A second definition of the mean is a sample-based definition. In this formulation, each sample point is typically given an equal weight

$$\bar{X} = \frac{1}{N} \sum_{i=1}^{N} x_i.$$

$$(2.59)$$

Table 2.1 presents a simulated sample of die rolls. Given these values, the sample mean becomes

$$\bar{X} = \frac{1}{10} \sum_{i=1}^{10} x_i = \frac{38}{10} = 3.8$$

$$(2.60)$$

which is close to the theoretical sample mean presented in Eq. (2.56).

The mean has several statistical applications, but intuitively the mean represents the central tendency of the random variable. However, the concept of the central tendency can be expanded past the value of the random variable itself. Specifically, the central tendency of a function which is defined as a transformation of the random variable can be defined:

Definition 2.14. The expected value of a function $g(x)$ can be expressed as $E[g(x)] = \int_{x \in X} g(x)f(x)dx$ where the integral is bounded.

Table 2.1.   Simulated sample of die rolls.

Observation	Value
1	4
2	6
3	5
4	5
5	2
6	5
7	1
8	2
9	6
10	2

In the study of the effect of risk on the agricultural firm, the analyst is frequently interested in the expected value of profit. Building on the triangular distribution function for the price of corn developed in the preceding section, the expected profit was defined as

$$E[\pi(p_y)] = p_y y - c \tag{2.61}$$

where $p_y \sim f(p_y)$ ($\sim$ denotes distributed as and $f(p_y)$ is the triangular probability distribution function), y is a fixed yield (this will be expanded to let y be another random variable below), and c is a fixed cost. Expected profit can then be derived as

$$E[\pi(p_y)] = \int_{1.75}^{3.25} [p_y y - c] f(p_y) dp_y = y \int_{1.75}^{3.25} p_y f(p_y) dp_y - c \tag{2.62}$$

since y and c are constants. Breaking the integral down into segments

$$E[\pi(p_y)] = y \left[\int_{1.75}^{2.30} p_y(a + bp_y) dp_y + \int_{2.30}^{3.25} p_y(c + dp_y) dp_y \right] - c$$

$$= y \left[\left(\frac{1}{2} a p_y^2 + \frac{1}{3} b p_y^3 \right) \Big|_{1.75}^{2.30} + \left(\frac{1}{2} c p_y^2 + \frac{1}{3} d p_y^3 \right) \Big|_{2.30}^{3.25} \right] - c$$

$$= y \times 2.433314 - c. \tag{2.63}$$

Substituting values for the output level and cost of production yields into Eq. (2.63) yields the expected profit level. This example is purely pedagogic in that profit is a linear function. Thus, from Eq. (2.62)

$$E[p_y] = \int_{1.75}^{3.25} p_y f(p_y) dp_y$$

$$\tag{2.64}$$

$$E[\pi(p_y)] = y E[p_y] - c$$

and

$$E[p_y] = \int_{1.75}^{2.30} p_y(a + bp_y) dp_y + \int_{2.30}^{3.25} p_y(c + dp_y) dp_y$$

$$= \left(\frac{1}{2} a p_y^2 + \frac{1}{3} b p_y^3 \right)\Big|_{1.75}^{2.30} + \left(\frac{1}{2} c p_y^2 + \frac{1}{3} d p_y^3 \right)\Big|_{2.30}^{3.25} = 2.433314.$$

$$\tag{2.65}$$

Extending the definition of expected values of functions a little further yields

Definition 2.15. The expected value of a bivariate function of random variables is defined as $E[h(x, y)] = \int_{y \in Y} \int_{x \in X} h(x, y) f(x, y) dx \, dy$ where $f(x, y)$ is a bivariate distribution function and the integral is bounded.

In the forgoing example, letting y be distributed uniformly $y \in [0, 1]$, under independence, the joint distribution function $(g(p_y, y))$ becomes

$$g(p_y, y) = f(p_y) \times 1. \tag{2.66}$$

Further because the random variables are independent

$$E[p_y y] = E[p_y] E[y]. \tag{2.67}$$

The expected value of the function is then defined by the result of Eq. (2.58) as

$$E[p_y y] = 2.433314 \times \frac{1}{2} = 1.216657. \tag{2.68}$$

Another frequently used function of random variables are the moments of the distribution. In general, the rth moment of a distribution is defined as

$$\mu_r(X) = E[X^r] = \int_{-\infty}^{\infty} x^r f(x)dx. \qquad (2.69)$$

Thus, the first moment of a distribution is equal to the mean

$$\mu_1(X) = E[X^1] = \int_{-\infty}^{\infty} x^r f(x)dx = E[X]. \qquad (2.70)$$

Again, using the uniform distribution as an example, the first four moments of the distribution are

$$\mu_1(X) = \int_0^1 x\,dx = \frac{1}{2}(x^2\big|_0^1 = \frac{1}{2}$$

$$\mu_2(X) = \int_0^1 x^2\,dx = \frac{1}{3}(x^3\big|_0^1 = \frac{1}{3}$$

$$\qquad\qquad (2.71)$$

$$\mu_3(X) = \int_0^1 x^3\,dx = \frac{1}{4}(x^4\big|_0^1 = \frac{1}{4}$$

$$\mu_4(X) = \int_0^1 x^4\,dx = \frac{1}{5}(x^5\big|_0^1 = \frac{1}{6}.$$

Another specification of the moments of a distribution is the central moments, or the moments about the mean of the distribution. The central moments of a distribution are defined as

$$\tilde{\mu}_r(X) = E[X - E[X]]^r = E[X - \mu_1(X)]. \qquad (2.72)$$

Most students have been exposed to several of the central moments. For example, the second central moment of a distribution is defined as the variance of the distribution

$$V[X] = E[X - E[X]]^2. \qquad (2.73)$$

The third central moment of the distribution is typically referred to as the skewness while the fourth central moment is the kurtosis.

In addition to univariate moments, the multivariate moments can also be defined similarly by the multidimensional integrals. The most frequently used multivariate central moment is the covariance between two random variables.

Definition 2.16. The *covariance* (or first central cross moment) between two random variables x and y is defined as $Cov(x,y) = E[(x - \mu_x)(y - \mu_y)]$ where μ_x and μ_y are the first moments or means of the respective random variable.

The covariance or first central cross moment will be particularly important in our discussion of diversification. Specifically, as long as the returns on two random variables are not perfectly correlated (see Eq. (2.28)), the overall level of risk can be reduced by holding a combination of the two investments.

Finally, associated with each distribution is a unique function called the moment generating function that can be used to derive the moments of that distribution.

Definition 2.17. The *moment generating function* $M_X(t)$ for the random variable x with distribution function $f(x)$ is defined as

$$M_X(t) = E[\exp[tx]] = \int_{-\infty}^{\infty} \exp[tx] f(x) dx \qquad (2.74)$$

such that this expectation exists in the neighborhood around $t = 0$.

If this moment generating function exists, the moments of the distribution are then defined by

$$\mu_r(x) = E[x^r] = M_X^{(r)}(0) = \left. \frac{d^r M_X(t)}{dt^r} \right|_{t=0}. \qquad (2.75)$$

As an example, consider the moment generating function for the univariate normal distribution

$$M_X(t) = \frac{1}{\sigma\sqrt{2\pi}} \int_{-\infty}^{\infty} \exp[tx] \exp\left[-\frac{(x - \mu)^2}{2\sigma^2}\right] dx$$

$$= \frac{1}{\sigma\sqrt{2\pi}} \int_{-\infty}^{\infty} \exp\left[\frac{2tx\sigma^2}{2\sigma^2} - \frac{x^2 - 2x\mu + \mu^2}{2\sigma^2}\right] dx$$

$$= \frac{1}{\sigma\sqrt{2\pi}} \int_{-\infty}^{\infty} \exp\left[\frac{-x^2 + 2tx\sigma^2 + 2x\mu - \mu^2}{2\sigma^2}\right] dx. \qquad (2.76)$$

Next, the exponent term can be solved by completing the square. Specifically, grouping the middle term in the quadratic in Eq. (2.76) yields $-x^2 + 2(t\sigma^2 + \mu)x - \mu^2$. To solve this expression, determine what expression has to be added, subtracted, or multiplied to make this expression a perfect square (or quadratic function). In this example,

$$(t\sigma^2 + \mu)^2 = t^2\sigma^4 + 2t\mu\sigma^2 + \mu^2. \tag{2.77}$$

Thus, adding and subtracting $t^2\sigma^4 + 2t\mu\sigma^2$ to the numerator results in

$$M_X(t) = \frac{1}{\sigma\sqrt{2\pi}}$$

$$\times \int_{-\infty}^{\infty} \exp\left[\frac{-x^2 + 2(t\sigma^2 + \mu)x - \mu^2 + t^2\sigma^4 + 2t\mu\sigma^2 - t^2\sigma^4 - 2t\mu\sigma^2}{2\sigma^2}\right] dx$$

$$= \frac{1}{\sigma\sqrt{2\pi}} \int_{-\infty}^{\infty} \exp\left[\frac{-x^2 + 2(t\sigma^2 + \mu)x - (t\sigma^2 + \mu)^2 + t^2\sigma^4 + 2t\mu\sigma^2}{2\sigma^2}\right] dx$$

$$= \frac{1}{\sigma\sqrt{2\pi}} \int_{-\infty}^{\infty} \exp\left[-\frac{(x - (t\sigma^2 + \mu))^2}{2\sigma^2} + \frac{1}{2}t^2\sigma^2 + t\mu\right] dx$$

$$= \exp\left[\frac{1}{2}t^2\sigma^2 + t\mu\right] \frac{1}{\sigma\sqrt{2\pi}} \int_{-\infty}^{\infty} \exp\left[-\frac{(x - (t\sigma^2 + \mu))^2}{2\sigma^2}\right] dx. \tag{2.78}$$

Since the last term in the final equation in Eq. (2.78) is simply the integral of a normal distribution with a different mean,

$$M_X(t) = \exp\left[\frac{1}{2}t^2\sigma^2 + t\mu\right]. \tag{2.79}$$

To demonstrate the usefulness of this representation, consider the first three moments of the normal distribution

$$M_X^{(1)}(0) = \frac{d\exp\left[\frac{1}{2}t^2\sigma^2 + t\mu\right]}{dt}\Bigg|_{t=0}$$

$$= (\sigma^2 t + \mu)\exp\left[\frac{1}{2}t^2\sigma^2 + t\mu\right]\Bigg|_{t=0} = \mu$$

$$M_X^{(2)}(0) = \left.\frac{d(\sigma^2 t + \mu)\exp\left[\frac{1}{2}t^2\sigma^2 + t\mu^2\right]}{dt}\right|_{t=0}$$

$$= \left.\left(\sigma^2 \exp\left[\frac{1}{2}t^2\sigma^2 + t\mu\right] + (\sigma^2 t + \mu)^2 \exp\left[\frac{1}{2}t^2\sigma^2 + t\mu\right]\right)\right|_{t=0}$$

$$= \sigma^2 + \mu^2$$

$$M_X^{(3)}(0) = \left.\frac{d\left(\sigma^2 \exp\left[\frac{1}{2}t^2\sigma^2 + t\mu\right] + (\sigma^2 t + \mu)^2 \exp\left[\frac{1}{2}t^2\sigma^2 + t\mu\right]\right)}{dt}\right|_{t=0}$$

$$= (\sigma^2(\sigma^2 t + \mu))\exp\left[\frac{1}{2}t^2\sigma^2 + t\mu\right]$$

$$+ 2(\sigma^2 t + \mu)\sigma^2 \exp\left[\frac{1}{2}t^2\sigma^2 + t\mu\right]$$

$$+ \left.(\sigma^2 t + \mu)^3 \exp\left[\frac{1}{2}t^2\sigma^2 + t\mu\right]\right|_{t=0}$$

$$= (3\sigma^2(\sigma^2 t + \mu))\exp\left[\frac{1}{2}t^2\sigma^2 + t\mu\right]$$

$$+ \left.(\sigma^2 t + \mu)^3 \exp\left[\frac{1}{2}t^2\sigma^2 + t\mu\right]\right|_{t=0}$$

$$= 3\sigma^2\mu + \mu^3. \tag{2.80}$$

2.6. Estimating Probability Functions

Knowledge of probability functions enters the analysis of the effect of risk and uncertainty on the agricultural firm on several different levels. At one level, the fact that the outcomes of economic decisions are affected by random variables such as prices, yields, and incomes define the scope of study. However, at a more applied level, most of the tools used to analyze decision making under risk require the researcher that parametrizes or estimates the factors determining the shape of the distribution function. This section provides a brief overview of the later procedures. Specifically, it begins by describing the two empirical procedures which can be used to elicit subjective probability density functions. This presentation is followed by two different objective procedures used to estimate the shape of probability density functions. The section concludes with a discussion of removing deterministic components from the distribution function such as time trends or production functions.

2.6.1. *Eliciting Subjective Probability Functions*

One of the vexing problems in modeling risk is the choice of subjective versus objective probabilities. Technically, a strong case can be made for the use of subjective distribution functions since the goal is to explain human responses to risk. However, little is known about the stability of subjective distribution functions. Further, it can be argued that if economic agents are rational that they will use the best information about random variables available to them. Thus, objective measures of probability distributions would at least form a lower-bound on the efficiency of producer's information. The later assumption is sometimes referred to as the rationality of decision makers. Abstracting away from the debate on the appropriateness of subjective distribution functions, this section describes two procedures for eliciting subjective probability functions.

The simplest approach is to elicit the triangular distribution function presented in Eq. (2.45). This approach amounts to asking the producer to specify the minimum, most likely, and maximum value of the random variable. This procedure is particularly adept at modeling continuous random variables such as output prices with which the decision maker is fairly familiar. However, the elicitation process does raise several efficiency problems. Notably, in its simplest form, the distribution function is unconditioned (i.e., does not depend on the possible values of other random variables). Other information could be imposed by scaling the distribution. For example, the research could provide the decision maker information about the current price of the commodity and the price of the commodity in the futures market.

A second approach is to directly elicit the relative probability. For example, the sample space for the price of corn could be divided into $0.25/bushel increments as presented in Table 2.2. Next, ask the decision maker to assign a number between 1 and 10 to each price which

Table 2.2. Direct elicitation of subjective probabilities.

Corn price	Relative merit	Probability
1.75	2	0.0488
2.00	8	0.1951
2.25	10	0.2439
2.50	9	0.2195
2.75	7	0.1707
3.00	6	0.1463
3.25	1	0.0244
Summation	41	1.0487

measures it relative merit or the relative likelihood of each market price. Summing up across the relative merit, and then dividing each relative merit by the sum yields the set of discrete probabilities in the last column of Table 2.2. A couple of caveats, first note that the probabilities in Table 2.2, are rounded to the fourth decimal place which introduces an error of 0.0487 which should be allocated across all the probabilities. Second, it may be worth allowing the decision maker to adjust the probabilities. Work out the implied mean and standard deviation. In this case, the probability weighted mean is 2.4754 and the probability weighted variance is 0.1244. Note the similarity between the two distributions. The expected value of the triangular distribution from Eq. (2.65) was 2.4333 and the variance for the triangular distribution is 0.0960.

2.6.2. *Objective Estimation of Probability Functions*

Typically the topic of estimating the parameters of a distribution function comprises entire courses. However, to facilitate analysis of the effect of risk on the agricultural firm, consider two general approaches to the estimation of the parameters of distribution functions: the method of moments and maximum likelihood.

The method of moment approach builds on the forgoing discussion of the moments of a distribution. Specifically, the method of moments solves for the parameters that equate the sample moments with the theoretical moments. Using the data presented in Table 2.3, the first and second moments of the sample can be computed. Next, assuming a distribution (e.g., the normal distribution), the researcher can solve for the parameters of the distribution by equating the sample moments with the theoretical moments

$$
\begin{aligned}
\mu_1(x) &= M_X^{(1)}(0) = \mu = 2.4123 \\
&\Rightarrow \hat{\mu}_{\text{MOM}} = 2.4123 \\
\mu_2(x) &= M_X^{(2)}(0) = \sigma^2 + \mu^2 = 5.9249 \\
&\Rightarrow \hat{\sigma}^2_{\text{MOM}} = 5.9249 - \hat{\mu}^2_{\text{MOM}} = 0.1060
\end{aligned}
\tag{2.81}
$$

where $\hat{\mu}_{\text{MOM}}$ denotes the mean of the distribution estimated using a method of moment estimator and $\hat{\sigma}^2_{\text{MOM}}$ denotes the method of moment estimator of the variance of the distribution.

Implicit in this formulation is the concept of a sample. Most statistical approaches assume that individuals are drawn from the sample consistent with the probability density function. For example, in this discussion of the cumulative density function, it has been demonstrated that

$$
z = F(x^*) = \int_{-\infty}^{x^*} f(x)dx \Rightarrow z \in [0, 1].
\tag{2.82}
$$

Given that $z = F(x^*)$ is a monotonically increasing function, the inverse mapping function from Eq. (2.40) can be defined as $x^* = F^{-1}(z)$. This inverse function is presented in Fig. 2.3, which actually

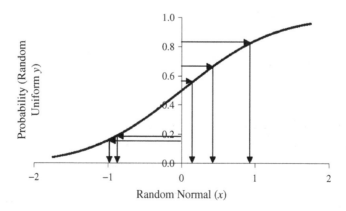

Fig. 2.3. Inverse mapping of a normal random variable into a uniform random variable.

maps a uniform distribution defined over the interval $[0, 1]$ and maps it onto a normal distribution function. In fact some random number generators use this procedure to generate normal random numbers. For this analysis, consider drawing five random numbers from the uniform distribution and map them using the inverse mapping in Fig. 2.3 into random normal space. For example, assume that the random draw is $\{0.1631, 0.1882, 0.5868, 0.6580, 0.8136\}$. Starting with the first point, $z = 0.1631 \Rightarrow x^* = -0.9820$. This inverse mapping is presented on Fig. 2.3. Moving to the second point, $z = 0.1882 \Rightarrow x^* = -0.8848$, which is mapped in the second set of arrows in Fig. 2.3. Continuing this mapping the vector $\{0.1631, 0.1882, 0.5868, 0.6580, 0.8136\}$ implies the normal variables $\{-0.9820, -0.8848, 0.2194, 0.4070, 0.8913\}$. Intuitively, the normal sample is "closer to the middle" than the uniform sample. In the standard normal, 66% of the probability mass lies within one standard deviation of the mean while in our transformed sample, 100% of the sample lies with one standard deviation of the mean (e.g., between -1 and 1).

The crux of the argument is that the random process generating the outcome is assumed to generate the sample in such a way that more likely events are generated relatively more often. Thus, the equally weighted sample is assumed to replicate the original distribution.

While this approach is relatively simple and tractable, it is not without its difficulties. For example, given the data and mean moments in Table 2.3, the parameters could be fitted with a gamma distribution

$$\mu_1(y) = \alpha\beta$$

$$\Rightarrow \alpha = \frac{2.4123}{\beta}$$

$$\mu_2(y) = \alpha(1 + \alpha)\beta^2$$

$$\Rightarrow 2.4123\left(1 + \frac{2.4123}{\beta}\right)\beta = 5.8192 + 2.4123\beta = 5.9249$$

$$\Rightarrow \hat{\beta}_{\text{MOM}} = 0.0438$$

$$\Rightarrow \hat{\alpha}_{\text{MOM}} = 55.0493. \tag{2.83}$$

Table 2.3.   Random values of corn price.

Observation number	Observe value	Squared value
1	2.4438	5.9721
2	2.5735	6.6227
3	1.9732	3.8934
4	2.1356	4.5609
5	2.7664	7.6531
6	2.2897	5.2429
7	2.3472	5.5093
8	2.5876	6.6956
9	2.0007	4.0029
10	2.6438	6.9896
11	2.3607	5.5729
12	2.4162	5.8378
13	2.4197	5.8547
14	2.6436	6.9885
15	2.9337	8.6066
16	1.9879	3.9518
17	3.0015	9.0092
18	2.5363	6.4330
19	2.0916	4.3749
20	2.6008	6.7643
21	1.8602	3.4604
22	1.7896	3.2027
23	2.8446	8.0920
24	2.6677	7.1167
25	2.3907	5.7153
Sample mean	2.4123	5.9249

Thus, the data could explain two different distributions. The question is then: which distribution is appropriate? One way to solve this question is to examine the third moment of each specification. The third empirical moment of the data presented in Table 2.3 is 14.7970. Working the third moment for each distribution from its moment generating function

$$\mu_3(x) = \mu^3 + 3\mu\sigma^2 \Rightarrow 14.8047$$

$$\text{such that } \hat{\mu}_{\text{MOM}} = 2.4123, \hat{\sigma}^2_{\text{MOM}} = 0.1060$$

$$\mu_3(z) = \alpha(1+\alpha)(2+\alpha)\beta^2 \Rightarrow 14.7909$$

$$\text{such that } \hat{\alpha}_{\text{MOM}} = 55.0493, \hat{\beta}_{\text{MOM}} = 0.0438.$$

(2.84)

So in both cases, the theoretical third moment of the distribution is close to the empirical moment (the same result holds for the fourth moment).

An alternative objective approach to estimating the parameters of a distribution function is by maximum likelihood. The argument behind maximum likelihood is to choose those parameters that maximize the likelihood or relative probability of drawing a particular sample. Building on the concept of independence, we assume that the sample presented in Table 2.3 was generated from *independently and identically distributed* (*iid*) draws from a normal distribution function. In this case, the independence assumption implies that we do not have to worry about correlation across the individual draws while the identical assumption means that each observation has the same mean (μ) and variance (σ^2). The likelihood function (or the probability of a particular sample) can then be written as

$$L = \prod_{i=1}^{N} f(x_i|\mu, \sigma^2) = (2\pi\sigma^2)^{-\frac{N}{2}} \exp\left[-\frac{1}{2\sigma^2} \sum_{i=1}^{N} (x_i - \mu)^2\right] \quad (2.85)$$

where L is the likelihood value. Maximizing this function with respect to the parameters μ and σ^2 implies

$$\max_{\mu, \sigma^2} L = (2\pi\sigma^2)^{-\frac{N}{2}} \exp\left[-\frac{1}{2\sigma^2} \sum_{i=1}^{N} (x_i - \mu)^2\right]. \quad (2.86)$$

Taking the first-order conditions with respect to μ first

$$\frac{\partial L}{\partial \mu} = (2\pi\sigma^2)^{-\frac{N}{2}} \exp\left[-\frac{1}{2\sigma^2}\sum_{i=1}^{N}(x_i - \mu)^2\right]$$

$$\times \left(-\frac{1}{2\sigma^2}\sum_{i=1}^{N}[-2(x_i - \mu)]\right) = 0$$

$$\Rightarrow \frac{1}{\sigma^2}\left(\sum_{i=1}^{N}x_i - N\mu\right) = 0 \Rightarrow \hat{\mu} = \frac{1}{N}\sum_{i=1}^{N}x_i. \qquad (2.87)$$

In order to solve for the first-order conditions with respect to the variance, treating σ^2 as a single variable

$$\frac{\partial L}{\partial \sigma^2} = -\frac{N}{2}(2\pi)^{-\frac{N}{2}}(\sigma^2)^{-\frac{N}{2}-1}\exp\left[-\frac{1}{2\sigma^2}\sum_{i=1}^{N}(x_i - \mu)^2\right]$$

$$+ (2\pi\sigma^2)^{-\frac{N}{2}}\left(\frac{(\sigma^2)^{-2}}{2}\sum_{i=1}^{N}(x_i - \mu)^2\right)$$

$$\times \exp\left[-\frac{1}{2\sigma^2}\sum_{i=1}^{N}(x_i - \mu)^2\right] = 0$$

$$\frac{\partial L}{\partial \sigma^2} = -\frac{N(\sigma^2)^{-1}}{2}(2\pi\sigma^2)^{-\frac{N}{2}}\exp\left[-\frac{1}{2\sigma^2}\sum_{i=1}^{N}(x_i - \mu)^2\right]$$

$$+ \left(\frac{(\sigma^2)^{-2}}{2}\sum_{i=1}^{N}(x_i - \mu)^2\right)(2\pi\sigma^2)^{-\frac{N}{2}}$$

$$\times \exp\left[-\frac{1}{2\sigma^2}\sum_{i=1}^{N}(x_i - \mu)^2\right]$$

$$= -\frac{N(\sigma^2)^{-1}}{2} + \frac{(\sigma^2)^{-2}}{2}\sum_{i=1}^{N}(x_i - \mu)^2$$

$$= -N\sigma^2 + \sum_{i=1}^{N}(x_i - \mu)^2 = 0 \Rightarrow \hat{\sigma}^2 = \frac{1}{N}\sum_{i=1}^{N}(x_i - \mu)^2.$$

$$(2.88)$$

The derivation of the maximum likelihood estimates as presented in Eqs. (2.89) and (2.88) can usually be simplified by maximizing the logarithm of the likelihood function. Taking the natural logarithm of the normal likelihood function

$$\ln(L) = -\frac{N}{2}\ln(\sigma^2) - \frac{1}{2\sigma^2}\sum_{i=1}^{N}(x_i - \mu)^2$$

$$\frac{\partial \ln(L)}{\partial \mu} = -\frac{1}{2\sigma^2}\sum_{i=1}^{N}[-2(x_i - \mu)] = 0$$

$$\Rightarrow \sum_{i=1}^{N} x_i - N\mu = 0 \Rightarrow \hat{\mu} = \frac{1}{N}\sum_{i=1}^{N} x_i$$

$$\frac{\partial \ln(L)}{\partial \sigma^2} = -\frac{N}{2}\frac{1}{\sigma^2} + \frac{1}{2(\sigma^2)^2}\sum_{i=1}^{N}(x_i - \mu)^2 = 0$$

$$\Rightarrow -N\sigma^2 + \sum_{i=1}^{N}(x_i - \mu)^2 = 0 \Rightarrow \hat{\sigma}^2 = \frac{1}{N}\sum_{i=1}^{N}(x_i - \mu)^2$$

$$(2.89)$$

which yields the same estimates as those obtained in Eqs. (2.87) and (2.88).

From the discussion of the method of moments estimators for the normal and comparing Eqs. (2.81) and (2.88) note that the method of moments estimator and maximum likelihood estimator of the parameters of the normal distribution are the same. The primary difference is the ability to make statistical inferences under the maximum likelihood estimator. Specifically, the variance of the estimated parameters obtain the *Cramer–Rao lower bound*

$$V(\hat{\theta}) \geq -\frac{1}{E\left[\frac{\partial^2 \ln(L)}{\partial \theta^2}\right]} \tag{2.90}$$

where θ are the parameters of the likelihood function.

2.6.3.   *Production Functions and Trends*

An additional complication in the estimation of distribution functions for use in decision making under risk and uncertainty is the effect of deterministic factors such as time trends and production functions which pervade economic analysis. For example, Fig. 2.4 presents the data for price received for tomatoes and the level of tomato production between 1960 and 2002 in Florida. Both series appear to have a significant upward trend. In fact, regressing the effect of time on both the price of tomatoes and the quantity of tomatoes produced yield parameter estimates that are significant at any conventional confidence level. To highlight the problem, first compute the parameters of the normal distribution without a trend line. The result is a mean price of \$22.28/cwt and a variance of 103.78, and a mean yield of 259.51 cwt/acre with a variance of 6187.45. Next, remove the trend for each data series by the result that $\hat{\varepsilon} = [I - x(x'x)x']y$.[1] The mean of each series is then zero under the assumptions of ordinary least squares, but in each case the variance is reduced. In the case of tomato prices, the variance falls to 18.13 while the variance in tomato yields falls to 1092.07. The reduction in variance is related to the explanatory power of the model. For example, the R^2 (or multiple correlation coefficient) is defined as the share of the variance explained by the regression. Using this definition, $1 - R^2$ is the remaining variation not explained by the regression. In this case, the R^2 for the regression on tomato prices (R_p^2) is 0.8252 while the R^2 for the regression on the tomato yields

---

[1] The standard result from regression analysis is that

$$\hat{\beta} = (x'x)^{-1}(x'y).$$

Substituting this result into the regression equation

$$\hat{\varepsilon} = y - x\beta$$
$$= y - x((x'x)^{-1}(x'y))$$
$$= y - x(x'x)^{-1}x'y = [I - x(x'x)x']y.$$

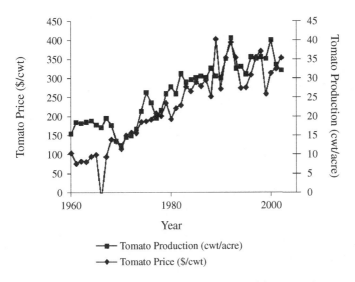

Fig. 2.4. Time series data of tomato prices and tomato yields in Florida, 1960–2002.

(R_Y^2) is 0.8235. Thus,

$$R_P^2 + \frac{18.13}{103.78} = 1$$

$$R_Y^2 + \frac{1092.07}{6187.45} = 1. \tag{2.91}$$

The question is then: what is the appropriate measure of risk? Subsequent chapters demonstrate that the variance parameter captures most of the disutility of risk in the normal distribution. Hence, the detrended yields appear to be less risky than the original data.

Another example can be seen in the estimation of production functions. For example, a traditional approach in agricultural economics is to model production using a parametric production function such as the Cobb–Douglas specification

$$y = x_1^\alpha x_2^\beta x_3^\gamma \tag{2.92}$$

where y is the level of outputs (such as tomatoes in the forgoing example), $x_i, i = 1, \ldots, 3$ are inputs (such as nitrogen, phosphorous, and potash in the case of crop production), and α, β, and γ are parameters.

To estimate the parameters in the Cobb–Douglas production function in Eq. (2.92), most researchers take the logarithm of both sides of the equation and append a constant and an error term yielding

$$\ln(y) = \alpha_0 + \alpha \ln(x_1) + \beta \ln(x_2) + \gamma \ln(x_3) + \varepsilon \qquad (2.93)$$

where α_0 is the constant and ε is regression error. The linear form of the Cobb–Douglas can then be estimated using a variety of techniques, but it is typically estimated using ordinary least squares. This procedure leads to several additional considerations in modeling the effect of risk on the agricultural firm. First, by the *Gauss–Markov theorem*, given the typical conditions on error term the parameter estimates are best linear unbiased. Note that the distributional assumptions required by Gauss–Markov only require that the individual errors be independently and identically distributed (or *homoscedasticity*); hence, the efficiency of the estimates is not dependent on the shape of the distribution function itself (as long as the distribution function does not violate homoscedasticity). Second, the distribution of the errors in Eq. (2.93) is not the distribution of risk in production space. Specifically, taking the exponential of each side of Eq. (2.93) yields

$$y = x_1^{\alpha} x_2^{\beta} x_3^{\gamma} e^{\alpha_0 + \varepsilon}. \qquad (2.94)$$

Hence, if $\varepsilon \sim N(0, \sigma_\varepsilon^2)$ the error in production space is distributed log-normally (or the log of the error is distributed normally). One difference is that while the normal distribution (or the transformed distribution) is symmetric, the log-normal distribution is positively skewed. The third central moments of the normal ($\mu_3^*(X)$) and log-normal ($\mu_3^*(Y)$) distribution are

$$\mu_3^*(X) = 0$$
$$\mu_3^*(Y) = \exp\left[3\mu + \frac{3\sigma^2}{2}\right](\exp[\sigma^2] - 1)^2(\exp[\sigma^2] + 2). \qquad (2.95)$$

Note that the third central moment of the log-normal distribution is always greater than zero. The fact that the skewness is limited for either distribution may limit the applicability of that distribution for our purposes.

Empirical Example: Table 2.4 presents the time-series sample of tomato prices and yields depicted in Fig. 2.4. We start by ignoring the possibility of a trend. As discussed above, this is the simplest case and likely to induce errors.

Computing the first empirical moment of the distribution along with the second, third, and fourth central moments for the random variables price (P) and yield (Υ) are given by

$$\begin{aligned}
\mu_1(P) &= 22.2800, & \mu_2^*(P) &= 103.7818, \\
\mu_3^*(P) &= -185.3151, & \mu_4^*(P) &= 53,367,955.4 \\
\mu_1(\Upsilon) &= 295.5116, & \mu_2^*(\Upsilon) &= 6,187.446, \\
\mu_3^*(\Upsilon) &= -35,174.9, & \mu_4^*(\Upsilon) &= 66,173,832.21.
\end{aligned} \quad (2.96)$$

Further, the covariance between the yield and price is 701.3036 implying a correlation coefficient of 0.8752. The results in Eq. (2.96) suggest that both distributions may be negatively skewed and kurtotic. One alternative for testing for skewness and kurtosis is the Bera–Jarque tests. First, normalize the third moment as

$$\alpha_3 = \frac{\mu_3^*(X)}{(\mu_2^*(X))^{\frac{3}{2}}} \Rightarrow \sqrt{T}\hat{\alpha}_3 \sim N(0,6). \quad (2.97)$$

Second, normalize the fourth moment of each distribution as

$$\alpha_4 = \frac{\mu_4^*(X)}{(\mu_2^*(X))^2} \Rightarrow \sqrt{T}(\hat{\alpha}_4 - 3) \sim N(0,24). \quad (2.98)$$

The joint test for skewness and kurtosis can then be written as

$$\frac{T}{6}(\hat{\alpha}_3)^2 + \frac{T}{24}(\hat{\alpha}_4 - 3)^2 \sim \chi_2^2. \quad (2.99)$$

The result is normalized skewness coefficients of -1.1359 for prices and -0.4684 for yields. In both cases, the results fail to reject the hypothesis that the distributions are not skewed. However, the normalized kurtosis coefficients are 38,128.7107 for prices and 11.2018 for yields. Thus, both empirical distributions appear to be more highly kurtotic than the normal (which has a kurtosis of 3) or leptokurtotic.

One alternative to the normal is the gamma distribution. As demonstrated above, the gamma distribution has a closed form solution for

Table 2.4.   Tomato prices and yields for Florida, 1960–2002.

Year	Tomato price ($/cwt)	Tomato production (cwt/acre)
1960	10.31	154
1961	7.60	184
1962	8.17	181
1963	8.02	185
1964	9.48	188
1965	9.93	177
1966	−2.10	170
1967	9.41	195
1968	13.95	176
1969	13.43	135
1970	11.42	124
1971	14.93	144
1972	15.71	149
1973	15.59	166
1974	18.54	213
1975	18.77	262
1976	19.16	235
1977	20.82	196
1978	20.07	215
1979	23.50	259
1980	19.28	277
1981	22.15	259
1982	22.90	311
1983	27.80	289
1984	26.60	295
1985	29.00	300
1986	27.90	305
1987	29.50	300
1988	25.20	325
1989	40.30	305
1990	27.20	300
1991	35.30	350
1992	39.40	405
1993	35.40	325
1994	27.40	330
1995	27.60	310
1996	30.90	355

(*Continued*)

Table 2.4. (*Continued*)

Year	Tomato price ($/cwt)	Tomato production (cwt/acre)
1997	35.50	350
1998	37.10	355
1999	25.90	350
2000	31.30	400
2001	32.40	335
2002	35.30	320

the method of moments approach. However, many of the distributions used in risk analysis do not have a closed form solution and, hence, they must be solved using numerical solution algorithms. In the current case, the parameters of the normal distribution can be solved for using the *Gauss–Sidel* approach which is explained more fully in Appendix C. Gauss–Sidel is an algorithm that is used to solve for the zeros of a multivariate nonlinear function. In the current example, the multivariate function is a vector function for the difference between the theoretical and empirical moments of a distribution

$$f(\alpha, \beta) = \begin{bmatrix} \alpha\beta - 22.2800 \\ \alpha\beta^2 - 103.7818 \end{bmatrix}. \tag{2.100}$$

This formulation solves for the values of α and β that yield a vector of zeros. To accomplish this, start with a first-order Taylor series expansion

$$f(\alpha, \beta) \approx f(\alpha_0, \beta_0) + \nabla_{\alpha,\beta} f(\alpha_0, \beta_0) \begin{bmatrix} \alpha - \alpha_0 \\ \beta - \beta_0 \end{bmatrix} \tag{2.101}$$

where α_0 and β_0 are initial guesses at the true value of α and β. Letting $f(\alpha, \beta) = 0$ and solving the system for α and β yields

$$-\nabla_{\alpha,\beta}^{-1} f(\alpha_0, \beta_0) f(\alpha_0, \beta_0) = \begin{bmatrix} \alpha - \alpha_0 \\ \beta - \beta_0 \end{bmatrix}$$

$$\Rightarrow \begin{bmatrix} \alpha \\ \beta \end{bmatrix} = \begin{bmatrix} \alpha_0 \\ \beta_0 \end{bmatrix} - \nabla_{\alpha,\beta}^{-1} f(\alpha_0, \beta_0) f(\alpha_0, \beta_0) \tag{2.102}$$

which results in a vector of α and β that are closer to the solution than α_0 and β_0. For the gamma distribution in this case

$$
\begin{bmatrix} \alpha \\ \beta \end{bmatrix} = \begin{bmatrix} \alpha_0 \\ \beta_0 \end{bmatrix} - \begin{bmatrix} \beta_0 & \alpha_0 \\ \beta_0^2 & 2\alpha_0\beta_0 \end{bmatrix}^{-1} \begin{bmatrix} \alpha_0\beta_0 - 22.2800 \\ \alpha_0\beta_0^2 - 103.7818 \end{bmatrix}
$$

$$
= \begin{bmatrix} \alpha_0 \\ \beta_0 \end{bmatrix} - \begin{bmatrix} \frac{2}{\beta_0} & -\frac{1}{\beta_0^2} \\ -\frac{1}{\alpha_0} & \frac{1}{\alpha_0\beta_0} \end{bmatrix} \begin{bmatrix} \alpha_0\beta_0 - 22.2800 \\ \alpha_0\beta_0^2 - 103.7818 \end{bmatrix}. \tag{2.103}
$$

Starting from an estimate of $\{\alpha_0, \beta_0\} = \{4, 4\}$ yields a value of $\{\alpha, \beta\} = \{4.65364, 4.916366\}$. Further within four iterations, the solution converges to $\{\alpha, \beta\} = \{4.7831, 4.65807\}$.

Next, to demonstrate the estimation of the likelihood functions using maximum likelihood, consider the formulation of the estimation problem for the gamma distribution for the same dataset, but including a trend line in the mean. Using the definition of the distribution function for the gamma distribution in Eq. (2.36), the likelihood function can be defined as

$$
L = \prod_{t=1}^{T} \frac{1}{\Gamma(\alpha)\beta^\alpha} x_t^{\alpha-1} \exp\left(-\frac{x_t}{\beta}\right)
$$

$$
\Rightarrow \ln(L) = \sum_{t=1}^{T} \left[-\ln(\Gamma\{\alpha\}) - \alpha\ln(\beta) + (\alpha-1)\ln(x_t) - \frac{x_t}{\beta} \right].
$$

$$\tag{2.104}$$

Next, redefine α to include the possibility of time-trend ($\alpha(t) = \alpha_0 + \alpha_1 t$). If $\alpha_1 = 0$, the results yield the same model estimated using the method of moments estimator above. Hence, the maximum likelihood approach involves maximizing

$$
\max_{\alpha_0, \alpha_1, \beta} \ln(L) = \sum_{t=1}^{T} \left[-\ln(\Gamma\{\alpha_0 + \alpha_1 t\}) - \{\alpha_0 + \alpha_1 t\}\ln(\beta) \right.
$$

$$
\left. + (\alpha_0 + \alpha_1 t - 1)\ln(x_t) - \frac{x_t}{\beta} \right]. \tag{2.105}
$$

Given the implicit nonlinearity involved in Eq. (2.105), solve for the optimum using the nonlinear optimization techniques presented in Appendix C. The gradient for this maximization problem becomes

$$
\nabla_{\alpha_0,\alpha_1,\beta} \ln(L) =
\begin{bmatrix}
-\sum_{t=1}^{T} \dfrac{\Gamma'(\alpha_0 + \alpha_1 t)}{\Gamma(\alpha_0 + \alpha_1 t)} - T \ln(\beta) + \sum_{t=1}^{N} \ln(x_t) \\[2ex]
-\sum_{t=1}^{T} \dfrac{\Gamma'(\alpha_0 + \alpha_1 t)}{\Gamma(\alpha_0 + \alpha_1 t)} t - \sum_{t=1}^{T} t \ln(\beta) + \sum_{t=1}^{T} t \ln(x_t) \\[2ex]
-\sum_{t=1}^{T} \dfrac{(\alpha_0 + \alpha_1 t)t}{\beta} + \dfrac{1}{\beta^2} \sum_{t=1}^{T} x_t
\end{bmatrix}'.
$$

$$(2.106)$$

The Hessian matrix then becomes

$$
\nabla^2_{\alpha_0,\alpha_1,\beta} \ln(L) =
\begin{bmatrix}
-\sum_{t=1}^{T}\left(\dfrac{\Gamma''(\alpha_0 + \alpha_1 t)}{\Gamma(\alpha_0 + \alpha_1 t)} - \left\{\dfrac{\Gamma'(\alpha_0 + \alpha_1 t)}{\Gamma(\alpha_0 + \alpha_1 t)}\right\}^2\right) & & \\[2ex]
-\sum_{t=1}^{T}\left(\dfrac{\Gamma''(\alpha_0 + \alpha_1 t)}{\Gamma(\alpha_0 + \alpha_1 t)} - \left\{\dfrac{\Gamma'(\alpha_0 + \alpha_1 t)}{\Gamma(\alpha_0 + \alpha_1 t)}\right\}^2\right)t & & \\[1ex]
-\dfrac{T}{\beta} & & \\[2ex]
-\sum_{t=1}^{T}\left(\dfrac{\Gamma''(\alpha_0 + \alpha_1 t)}{\Gamma(\alpha_0 + \alpha_1 t)} - \left\{\dfrac{\Gamma'(\alpha_0 + \alpha_1 t)}{\Gamma(\alpha_0 + \alpha_1 t)}\right\}^2\right)t & & -\dfrac{T}{\beta} \\[2ex]
-\sum_{t=1}^{T}\left(\dfrac{\Gamma''(\alpha_0 + \alpha_1 t)}{\Gamma(\alpha_0 + \alpha_1 t)} - \left\{\dfrac{\Gamma'(\alpha_0 + \alpha_1 t)}{\Gamma(\alpha_0 + \alpha_1 t)}\right\}^2\right)t^2 & & -\dfrac{1}{\beta}\sum_{t=1}^{T}t \\[2ex]
-\dfrac{1}{\beta}\sum_{t=1}^{T}t & & \sum_{t=1}^{T}\left\{\dfrac{\alpha_0 + \alpha_1 t}{\beta^2} - \dfrac{2}{\beta}x_t\right\}
\end{bmatrix}.
$$

$$(2.107)$$

In both the gradient and the Hessian, the derivatives of the gamma function are numerical. To apply this procedure requires a starting guess of the parameters. Here, start with the ordinary least squares estimates of $\alpha_0 = 131.436$ and $\alpha_1 = 5.8900$, and an estimate of $\beta = 0.50$. Given these values $\ln(L) = -3,581.32$. The numerical value of the gradient is

$$\nabla_{\alpha_0,\alpha_1,\beta} \ln(L) = [28.822 \quad 618.741 \quad 21678.0] \tag{2.108}$$

and the numerical value of the Hessian is

$$\nabla^2_{\alpha_0,\alpha_1,\beta} \ln(L) = \begin{bmatrix} -0.177667 & -3.18113 & -84.0 \\ -3.18113 & -82.593 & -1806.0 \\ -84.0 & -1806.0 & -130068 \end{bmatrix} \tag{2.109}$$

which implies $\alpha_0 = 201.302$, $\alpha_1 = 8.96697$, and $\beta = 0.578822$. The new value of the log-likelihood function is $\ln(L) = -431.255$. Reapplying the Newton–Raphson steps 50 times yields the optimal parameter estimates $\alpha_0 = 27.6558$, $\alpha_1 = 1.20804$, and $\beta = 4.81219$.

2.7. Martingales and Random Walks

As a final topic in the discussion of probability, consider the *random* formulation. In this discussion, the beginning point is the development of Martingales. As with most of the models of risk, Martingales have their basis in gambling concepts. Suppose that we have a sequence of random variables $X_t : t = 1, \dots$ (or $X_1, X_2, \dots$) defined on a measure space (C, B, P) (i.e., where C is the sample space, B is the Borel set, and P is the probability measure in Definition 2.9). Let the random variable be the total winnings for the gambler after n games of chance $(X_i : i = 1, \dots, n)$. The question is then: what will the return be on the next game X_{n+1}? If the game is fair, the gambler will neither be richer or poorer after the next game. Specifically, if the game is fair $E[X_{n+1}] = 0$ (also, the expected winnings on all the previous games is also zero). Thus, if the game is fair, this result is independent of past winnings. Mathematically,

$$E[X_{n+1}|X_1, \dots, X_n] = X_n \tag{2.110}$$

(i.e., the wealth after the bet is the same as the wealth before the bet).

Definition 2.18. The sequence $\{(X_n, B_n), n = 1, 2, \dots\}$ is a *Martingale* if each n satisfies

(1) $C_n \subset C_{n+1}$ or the Borel sets are nested. Thus, the set containing the current draw are contained in the set containing future draws,

(2) X_n is measurable on C_n or the sequence of random variables are defined on the nested Borel sets,

(3) $E[|X_n|] < \infty$ or each expectation is bounded, and

(4) $E[X_{n+1}|C_n] = X_n$.

As an example, consider Schmitz's (1995) discussion of farmland values. Schmitz starts by defining the price of farmland as the present value of the future expected cash flows arising from that land

$$V_t = \sum_{s=1}^{\infty} \frac{E_t[CF_{t+s}|\Omega_t]}{(1+r)^s}$$

$$= \frac{E_t[CF_{t+1}|\Omega_t]}{(1+r)} + \sum_{s=2}^{\infty} \frac{E_t[CF_{t+s}|\Omega_t]}{(1+r)^s} \qquad (2.111)$$

where V_t is the value of farmland, $E[CF_{t+s}|\Omega_t]$ is the expected value of the cash flow from the farmland in period $t+s$ (CF_{t+s}) given the information set available in period t (Ω_t), and r is the discount rate. Abstracting away from the expectations operator momentarily, rewrite Eq. (2.111) as

$$V_t = \frac{CF_{t+1}}{(1+r)} + \frac{1}{(1+r)} \sum_{s=2}^{\infty} \frac{CF_{t+s}}{(1+r)^s}$$

$$= \frac{CF_{t+1}}{(1+r)} + \frac{1}{(1+r)} V_{t+1}. \qquad (2.112)$$

Solving Eq. (2.112) yields

$$V_t - V_{t+1} = \frac{CF_{t+1}}{(1+r)} - rV_t = \varepsilon_t \qquad (2.113)$$

so that if $CF_t/(1+r) = rV_t$ there is no change in farmland values.

Next, add the random nature of cash flows back into the discussion, but instead of using Schmitz's notation for the information set, adopt the nested Borel sets as described in Definition 2.18 (also see Malliaris and Brock, 1982)

$$v_t = E[V_t|C_t]$$

$$= E\left[\sum_{s=1}^{\infty} \frac{CF_{t+s}}{(1+r)^s}\middle| C_t\right] \qquad (2.114)$$

$$= \sum_{s=1}^{\infty} E\left[\frac{CF_{t+s}}{(1+r)^s}\middle| C_t\right]$$

by the linearity of the expectation. Next, walk the valuation ahead one period in a similar fashion as in Eq. (2.112)

$$E[v_{t+1}|C_t] = E\left[E\left[\sum_{s=2}^{\infty} \frac{CF_{t+s}}{(1+r)^{s-1}}\middle| C_{t+1}\right]\middle| C_t\right] \qquad (2.115)$$

since the present value of the land in the next period $(t+1)$ is the present value of cash flows starting in period $t+2$ and going forward. Thus, the cash flow for period CF_{t+2} (in the value v_{t+1}) is discounted by a discount factor of $(1+r) \Leftarrow (1+r)^{s-1}|_{s=2} \Leftarrow (1+r)^{2-1}$. Next, since $C_t \subset C_{t+1}$, hence, by Definition 2.18(4)

$$E\left[E\left[\sum_{s=2}^{\infty} \frac{CF_{t+s}}{(1+r)^{s-1}}\middle| C_{t+1}\right]\middle| C_t\right] = E\left[\sum_{s=2}^{\infty} \frac{CF_{t+s}}{(1+r)^{s-1}}\middle| C_t\right].$$
$$\qquad (2.116)$$

Next, add and subtract CF_{t+1} yielding

$$E[v_{t+1}|C_t] = E\left[\left(CF_t - CF_t + \sum_{s=2}^{\infty} \frac{CF_{t+s}}{(1+r)^{s-1}}\right)\middle| C_t\right]$$

$$= E\left[\left(CF_t + \sum_{s=2}^{\infty} \frac{CF_{t+s}}{(1+r)^{s-1}}\right)\middle| C_t\right]$$

$$\times \frac{(1+r)}{(1+r)} - E[CF_t|C_t]$$

$$= E\left[\left(\frac{CF_t}{(1+r)} + \sum_{s=2}^{\infty} \frac{CF_{t+s}}{(1+r)^s}\right)\bigg| C_t\right](1+r) - E[CF_t|C_t]$$

$$= E\left[\sum_{s=1}^{\infty} \frac{CF_t}{(1+r)^s}\bigg| C_t\right](1+r) - E[CF_t|C_t].$$

$$(2.117)$$

Given the result in Eq. (2.114), the last expression becomes

$$E[v_{t+1}|C_t] = (1+r)v_t - E[CF_{t+1}|C_t]. \tag{2.118}$$

The result is a stochastic version of Eq. (2.113), or that farmland values follow a Martingale if

$$E[v_{t+1}|C_t] - v_t = rv_t - E[CF_{t+1}|C_t]. \tag{2.119}$$

Building on this formulation, the set of random variables described above can be rewritten as $\{X_t, t \in T\} \Rightarrow X_{t_i}(\omega) : t_1, t_2, \ldots, t_n \in T$ or that a set of random variables building on the measure theory introduced earlier in this chapter. Thus, the probability of a given sequence of random variables can be defined as

$$f(X_{t_1}(\omega), X_{t_2}(\omega), \ldots, X_{t_n}(\omega))$$
$$= P[\omega : X_{t_1}(\omega), X_{t_2}(\omega), \ldots, X_{t_n}(\omega) \in \Omega]. \tag{2.120}$$

The most useful random process from may be is the Wiener process or the Brownian motion process defined as $\{z_t, t \in [0, \infty)\}$ defined on a probability space (Ω, C, P) with the properties $z_0(\omega) = 0$. For

$$P[z_{t_i} - z_{t_{i=1}} \in H_i \text{ for } i \leq n] = \prod_{i \leq n} P[z_{t_i} - z_{t_{i=1}} \in H_i] \tag{2.121}$$

or the increments are independent. For any increment in time (for $t > s \geq 0$)

$$P[z_t - z_s \in H] = \frac{1}{\sqrt{2\pi(t-s)}} \int_H \exp\left[-\frac{1}{2}\frac{x^2}{(t-s)}\right] ds \tag{2.122}$$

or that the increments between any two points of the process are normally distributed. And, finally, for each $\omega \in \Omega$, $z_t(\omega)$ is continuous at t.

Departing from the measure definition of a Wiener process in Eqs. (2.121) and (2.122), the Wiener process is typically written as a random walk

$$X_t = X_{t-1} + \varepsilon_t, \varepsilon_t \sim N(0,1). \tag{2.123}$$

Following the definition in Eq. (1.122)

$$X_t - X_{t-1} = \varepsilon_t \Rightarrow f(\varepsilon_t) = \frac{1}{\sqrt{2\pi}} \exp\left[-\frac{1}{2}\varepsilon_t^2\right] \tag{2.124}$$

so X_t is consistent with the definition of a Wiener process.

2.8.   Summary

This chapter develops two important concepts in risk analysis. First, this chapter provides a fairly rigorous definition of random events. Second, the chapter demonstrates several different ways to estimate the parameters which describe risk.

Chapter 3

Expected Utility — The Economic Basis of Decision Making Under Risk

At the basic level, any scientific or disciplinary pursuit attempts to develop a systematic framework for the analysis of a physical or human phenomenon. In the case of economics, the phenomenon studied is the way humans allocate resources to meet unlimited and competing human wants and desires. Given this overall goal, most disciplines formulate a core theory which represents the accepted body of knowledge regarding the physical or human phenomenon. This core body serves as the source of new hypotheses, which are conjectures or new theory which will be tested using empirical data or sometimes mathematical deduction. In some cases, after repeated verification, the science may choose to elevate a particular theory to the status of a law. This elevation is a somewhat mystical process which is relatively rare in economics (i.e., economics has fairly few facts which it considers economic laws).

This chapter presents one of the more abstract and often contentious theories in economics, the expected utility hypothesis. The expected utility hypothesis was developed as an axiomatic proof by von Neumann and Morgenstern (1953), which is a logical proof based on deductions from a set of economic axioms. From a mathematical perspective, if the axioms are correct, the conclusion must follow. However, for all of its elegance, the expected utility hypothesis has numerous detractors. Most of these detractors emphasize the possibility of anomalous behavior that contradicts the conclusions of the hypothesis.

As a starting point, this chapter redevelops several of the dual consumption relationships found in standard microeconomic textbooks. Specifically, the chapter derives the duality between the *indirect utility function* and the *expenditure function*. This duality is important because the indirect utility function is similar to the preference mapping relationship derived from expected utility. The duality between the indirect utility and the expenditure function is then useful in the derivation of the *certainty equivalent* (which is defined as the certain amount that a decision maker is willing to give or take for the risky alternative) and the risk premium (which is defined as the maximum price that the decision maker is willing to pay to forgo the risk). Implicitly while most of this book examines the producers' decisions, the risk preferences are typically developed from the standpoint of consumption preferences.

3.1.   Consumption and Utility

As a starting point, consider the reduction of the standard utility maximization problem to the utility of income. This reduction allows the analyst to abstract away from the effect of volatility on the consumption of individual commodities to the effect of uncertainty on consumer well-being in general.

A typical economic assumption is that economic agents (consumers, producers, etc.) behave in a way that maximizes their expected utility. The typical formulation is

$$\max_{x_1,x_2} U(x_1,x_2)$$
$$st \; p_1 x_1 + p_2 x_2 \leq \Upsilon \tag{3.1}$$

where x_1 and x_2 are the consumption goods, p_1 is the price of good 1, p_2 is the price of good 2, and Υ is the monetary income. In the development of decision making under risk, the consumption goods and the prices for those goods are typically assumed to be known with certainty (an alternative where the goods themselves are risky can be found in Chapter 16 of Schmitz *et al.*, where there is a health risk associated with the consumption of goods from different countries). Income is assumed to be random. Thus, the researcher is less interested in the tradeoffs between goods than the effect of the overall uncertainty of income on the well-being of the consumer. Hence, utility is redefined as a function of income given that all the optimal consumption decisions have been made.

The linkage between these two concepts is the indirect utility function, which solves for the optimizing behavior by the economic agent as a function of the prices of the consumption goods and income. Specifically, using the Cobb–Douglas utility function, the utility maximization problem can be rewritten as

$$\max_{x_1, x_2} x_1^{\alpha} x_2^{\beta}$$
$$\text{st } p_1 x_1 + p_2 x_2 \leq \Upsilon. \tag{3.2}$$

Since the utility function is strictly increasing, the inequality can be replaced with an equality. The maximization problem can then be reformulated as a Lagrangian:

$$L = x_1^{\alpha} x_2^{\beta} + \lambda(\Upsilon - p_1 x_1 - p_2 x_2). \tag{3.3}$$

The first-order conditions are then:

$$\frac{\partial L}{\partial x_1} = \alpha \frac{x_1^{\alpha} x_2^{\beta}}{x_1} - \lambda p_1 = 0$$

$$\frac{\partial L}{\partial x_2} = \beta \frac{x_1^{\alpha} x_2^{\beta}}{x_2} - \lambda p_2 = 0 \tag{3.4}$$

$$\frac{\partial L}{\partial \lambda} = \Upsilon - p_1 x_1 - p_2 x_2 = 0.$$

Taking the ratio of the first two first-order conditions in Eq. (3.4) yields

$$x_2 = \frac{\beta}{\alpha} \frac{x_1 p_1}{p_2}. \tag{3.5}$$

Substituting the result from Eq. (3.5) into the third first-order condition from Eq. (3.4) yields the demand for x_1 as a function of prices and income

$$x_1(p_1, p_2, \Upsilon) = \frac{\Upsilon}{p_1} \left(\frac{\alpha}{\alpha + \beta} \right). \tag{3.6}$$

Substituting the relationship for x_1 in Eq. (3.6) into the relationship between x_1 and x_2 in Eq. (3.7) yields the *Marshallian demand* for x_2

$$x_2(p_1, p_2, \Upsilon) = \frac{\Upsilon}{p_2}\left(\frac{\beta}{\alpha + \beta}\right). \qquad (3.7)$$

Each Marshallian demand curve (in Eqs. (3.6) and (3.7)) are substituted into the original utility function in Eq. (3.2) to yield the indirect utility function

$$V(p_1, p_2, \Upsilon) = \frac{\Upsilon^{\alpha+\beta}}{(\alpha + \beta)^{\alpha+\beta}}\left(\frac{\alpha}{p_1}\right)^{\alpha}\left(\frac{\beta}{p_2}\right)^{\beta}. \qquad (3.8)$$

This indirect utility function embodies the optimizing behavior from each Marshallian demand function. Thus, the indirect utility function relates the utility directly to income and prices assuming the optimizing behavior.

The expenditure function, as the indirect utility function, examines the implications of optimizing behavior. However, the expenditure function determines the minimum income required to provide a given level of utility. Mathematically, the expenditure function is given by the solution of

$$\min_{x_1, x_2} p_1 x_1 + p_2 x_2$$

$$st \ U(x_1, x_2) \geq U^*, \qquad (3.9)$$

where U^* is some required level of utility. Using the same Cobb–Douglas utility function described earlier, this minimization problem becomes

$$\min_{x_1, x_2} p_1 x_1 + p_2 x_2$$

$$st \ x_1^{\alpha} x_2^{\beta} = U^*. \qquad (3.10)$$

The Lagrangian for this problem becomes

$$L = p_1 x_1 + p_2 x_2 + \mu(U^* - x_1^{\alpha} x_2^{\beta}). \qquad (3.11)$$

The first-order conditions for the minimization problem specified in Eq. (3.11) are then

$$\frac{\partial L}{\partial x_1} = p_1 - \mu\alpha\frac{x_1^\alpha x_2^\beta}{x_1} = 0$$

$$\frac{\partial L}{\partial x_2} = p_2 - \mu\beta\frac{x_1^\alpha x_2^\beta}{x_2} = 0 \qquad (3.12)$$

$$\frac{\partial L}{\partial \mu} = U^* - x_1^\alpha x_2^\beta = 0.$$

Again, taking the ratio of the first two first-order conditions in Eq. (3.12) yields

$$x_2 = \frac{p_1\beta}{p_2\alpha}. \qquad (3.13)$$

Substituting the result from Eq. (3.13) into the third first-order condition of Eq. (3.12) yields

$$U^* - x_1^\alpha\left(\frac{p_1\beta}{p_2\alpha}x_1\right)^\beta = 0$$

$$U^* - x_1^{\alpha+\beta}\left(\frac{p_1\beta}{p_2\alpha}\right)^\beta = 0 \qquad (3.14)$$

$$x_1^{\alpha+\beta} = U^*\left(\frac{p_2\alpha}{p_1\beta}\right)^\beta.$$

The optimum level of x_1 given a fixed level of utility and prices (the *Hicksian demand* curve) is then

$$x_1^h(p_1, p_2, U) = U^{\frac{1}{\alpha+\beta}}\left(\frac{p_2\alpha}{p_1\beta}\right)^{\frac{\beta}{\alpha+\beta}}. \qquad (3.15)$$

Similarly, the Hicksian demand for x_2 (derived by substituting the Hicksian demand curve for x_1 into the ratio derived from the first-order conditions in Eq. (3.13)) is

$$x_2^h(p_1, p_2, U) = U^{\frac{1}{\alpha+\beta}}\left(\frac{p_1\beta}{p_2\alpha}\right)^{\frac{\alpha}{\alpha+\beta}}. \qquad (3.16)$$

Following the same approach used to define the indirect utility function, each Hicksian demand curve (from Eqs. (3.15) and (3.16)) is substituted back into the consumer's budget function in Eq. (3.10) to derive the expenditure function

$$
\begin{aligned}
e(p_1, p_2, U) &= p_1 U^{\frac{1}{\alpha+\beta}} p_1^{\frac{-\beta}{\alpha+\beta}} p_2^{\frac{\beta}{\alpha+\beta}} \alpha^{\frac{\beta}{\alpha+\beta}} \beta^{\frac{-\beta}{\alpha+\beta}} \\
&\quad + p_2 U^{\frac{1}{\alpha+\beta}} p_1^{\frac{\alpha}{\alpha+\beta}} p_2^{\frac{-\alpha}{\alpha+\beta}} \alpha^{\frac{-\alpha}{\alpha+\beta}} \beta^{\frac{\alpha}{\alpha+\beta}} \\
&= p_1^{\frac{\alpha}{\alpha+\beta}} U^{\frac{1}{\alpha+\beta}} p_2^{\frac{\beta}{\alpha+\beta}} \alpha^{\frac{\beta}{\alpha+\beta}} \beta^{\frac{-\beta}{\alpha+\beta}} \\
&\quad + p_2^{\frac{\beta}{\alpha+\beta}} U^{\frac{1}{\alpha+\beta}} p_1^{\frac{\alpha}{\alpha+\beta}} \alpha^{\frac{-\alpha}{\alpha+\beta}} \beta^{\frac{\alpha}{\alpha+\beta}} \\
&= U^{\frac{1}{\alpha+\beta}} p_1^{\frac{\alpha}{\alpha+\beta}} p_2^{\frac{\beta}{\alpha+\beta}} \alpha^{\frac{-\alpha}{\alpha+\beta}} \beta^{\frac{-\beta}{\alpha+\beta}} (\alpha + \beta).
\end{aligned}
\tag{3.17}
$$

Note the duality between the functions. Starting with the expenditure function

$$
\begin{aligned}
e(p_1, p_2, U) &= U^{\frac{1}{\alpha+\beta}} p_1^{\frac{\alpha}{\alpha+\beta}} p_2^{\frac{\beta}{\alpha+\beta}} \alpha^{\frac{-\alpha}{\alpha+\beta}} \beta^{\frac{-\beta}{\alpha+\beta}} (\alpha + \beta) = \Upsilon \\
U p_1^{\alpha} p_2^{\beta} \alpha^{-\alpha} \beta^{-\beta} (\alpha + \beta)^{\alpha+\beta} &= \Upsilon^{\alpha+\beta} \\
U &= p_1^{-\alpha} p_2^{-\beta} \alpha^{\alpha} \beta^{\beta} (\alpha + \beta)^{\alpha+\beta} \Upsilon^{\alpha+\beta} \\
V(p_1, p_2, \Upsilon) &= \left(\frac{\alpha}{p_1}\right)^{\alpha} \left(\frac{\beta}{p_2}\right)^{\beta} \frac{\Upsilon^{\alpha+\beta}}{(\alpha + \beta)^{\alpha+\beta}}.
\end{aligned}
\tag{3.18}
$$

The relationship between the expenditure function and the indirect utility function can be used to prove several useful results for demand theory. For example, substituting the expenditure function into the income term of the Marshallian demand curve yields the Slutsky decomposition of demand. Substituting the expenditure function into the indirect utility function, holding the level of utility constant, and differentiating the result with respect to prices yields Roy's identity. However, for the purpose of analyzing risk, the relationship between the maximum level of utility (as manifested by the indirect utility function) and the minimum level of expenditure required to generate that level of utility is extremely useful. This relationship implies that the indirect utility function is a monotonically increasing function of income.

Hence, the relationship between income and utility is a one-to-one relationship. Given this one-to-one relationship, the function can be inverted to derive that level of income required to generate any given level of utility.

3.2. Expected Utility

Risk analysis combines this one-to-one mapping along with an axiomatic proof to derive the economic value of a risky alternative to decision makers. Before introducing the axiomatic proof, consider how the expected value of the indirect utility function can be used along with the expenditure function to derive the consequences of risk. First, assume that producers behave in a manner that maximizes their expected utility. Implicitly, this says that a producer is indifferent between two risky alternatives that yield the same expected utility. The *certainty equivalent* is defined as the certain amount that an individual is willing to accept instead of the risky investment. Specifically, letting one of the alternatives be a risk-free alternative (action a_0), if the economic agent is indifferent between the risk-free alternative and a risky alternative (action a_1) action, a_0 is the certainty equivalent of a_1. Alternatively, the utility from the risk-free action equals the expected utility from the risky action.

Simplifying the indirect utility function in Eq. (3.8) by ignoring the price of the consumption goods yields a *power utility function*:

$$U(\Upsilon) = \frac{\Upsilon^{1-r}}{1-r}, \tag{3.19}$$

where r is a parameter that will be shown in Chapter 4 to control risk preferences. Using the power utility function, the *expected utility* of the risky alternative is depicted as some point on the arc below the utility function depicted in Fig. 3.1. To demonstrate the use of the indirect utility function and the expenditure function, start by assuming that a risky investment has a Bernoulli distribution paying $150,000 with probability 0.6 and $50,000 with probability 0.4. The economic question then could be: What is this investment alternative worth? Assume $r = 0.5$, then the expected utility of the

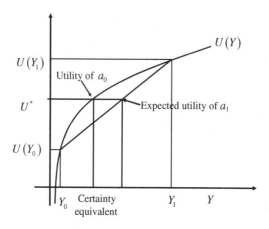

Fig. 3.1.   The indirect utility function and expected utility.

gamble is:

$$E[U(\Upsilon)] = 0.6\frac{(150,000)^{1-0.5}}{1-0.5} + 0.4\frac{(50,000)^{1-0.5}}{1-0.5} = 643.64.$$

(3.20)

The certainty equivalent (or certain amount that the decision maker is willing to take or give for this opportunity) of the gamble can be computed as

$$U = \frac{\Upsilon^{1-r}}{1-r}$$

$$(1-r)U = \Upsilon^{1-r}$$

$$[(1-r)U]^{\frac{1}{1-r}} = \Upsilon$$

$$[(0.5)643.64]^{\frac{1}{0.5}} = \Upsilon = 103,569.20.$$

(3.21)

Note that the expected value of the gamble is \$110,000. This implies a *risk premium* (the amount that an individual would be willing to pay to forgo the risky gamble) of $110,000 - 103,569.20 = \$6,430.78$.

Working on a slightly different problem, assume that the risky gamble that pays \$150,000 with some probability p and \$50,000 with probability $(1 - p)$. In this formulation, p can be derived that makes the decision maker indifferent between the risky gamble and a certain payoff of \$108,000. Naturally, p is higher than 0.6. Using the power utility function,

$$U(108,000) = \frac{108,000^{0.5}}{0.5} = 657.27. \qquad (3.22)$$

The probability which equates the expected utility to 657.27 is

$$657.27 = 774.60\, p + 447.21(1-p)$$
$$657.27 = 774.50\, p + 447.21 - 447.21p$$
$$210.05 = 327.38p \qquad (3.23)$$
$$p = 0.6416.$$

Changing the problem slightly, assume that the payoffs are \$150,000 with probability 0.6 and \$50,000 with probability 0.4. What is the r required to make the certainty equivalent \$108,000? The conjecture is that this risk aversion is less than the original risk aversion of 0.5. Borrowing from the above analysis, the problem this time is slightly different

$$\frac{108,000^{1-r}}{1-r} = 0.6\frac{150,000^{1-r}}{1-r} + 0.4\frac{50,000^{1-r}}{1-r}. \qquad (3.24)$$

As long as r does not equal 1, Eq. (3.24) can be simplified and written in an implicit functional form as:

$$g(r) = 108,000^{1-r} - 0.6 \cdot 150,000^{1-r} - 0.4 \cdot 50,000^{1-r} = 0.$$
$$(3.25)$$

The solution of r becomes 3.65, leading to a $1 - r$ of -2.65 (the solution is based on Gauss–Seidel as described in Appendix C).

3.2.1. *Intuition behind Von Neumann and Morgenstern Proof*

Von Neumann and Morgenstern offer an axiomatic proof to the conjecture that economic agents prefer alternatives that yield higher expected utility based on the first line of reasoning. Specifically, using the forgoing example consider two risky alternatives. The first gamble produces an outcome of \$75,000 with a probability of 0.4 and \$125,000 with a probability of 0.6 while the second gamble pays \$80,000 with a probability of 0.5 and \$120,000 with a probability of 0.5. In each case, a probability can be derived using the approach in Eq. (3.20), which makes each of these prospects equally preferable to (or indifferent to) the original gamble (i.e., a gamble that pays \$50,000 with some probability and \$150,000 with another probability).

Solving for the probabilities of the first alternative

$$\frac{125,000^{0.5}}{0.5}(0.4) + \frac{75,000^{0.5}}{0.5}(0.6) = 643.3531$$

$$\Rightarrow 643.3531 = \frac{150,000^{0.5}}{0.5}\alpha_1 + \frac{50,000^{0.5}}{0.5}(1 - \alpha_1)$$

$$= 774.5967\alpha_1 + 447.2136(1 - \alpha_1) \Rightarrow \alpha_1 = 0.5991. \quad (3.26)$$

Thus, the gamble which pays \$75,000 with probability 0.4 and \$125,000 with probability 0.6 is indifferent to a gamble that pays \$50,000 with probability 0.4009 and \$150,000 with probability 0.5991.

Solving for the probabilities for the gamble in Eq. (3.20) that make the second gamble equivalent to the index gamble

$$\frac{120,000^{0.5}}{0.5}(0.5) + \frac{80,000^{0.5}}{0.5}(0.50) = 629.2529$$

$$\Rightarrow 629.2529 = \frac{150,000^{0.5}}{0.5}\alpha_2 + \frac{50,000^{0.5}}{0.5}(1 - \alpha_2)$$

$$= 774.5967\alpha_2 + 447.2136(1 - \alpha_2) \Rightarrow \alpha_2 = 0.5560. \quad (3.27)$$

Note that $\alpha_1 > \alpha_2$. This inequality has implications for the certainty equivalents. Taking the inverse of the expected utility value in Eq. (3.26) yields a certainty equivalent of \$103,475.80 while the inverse of the expected utility value in Eq. (3.27) is \$98,989.79. By the certainty equivalents, the decision maker prefers the first gamble to the second gamble (because the first gamble has larger certainty equivalence). However, the same conclusion could be reached based on the probabilities. In fact, the probability approach is actually the approach used in the von Neumann and Morgenstern proof.

The conjectures under von Neumann and Morgenstern assume that you have three points $A, B,$ and C (depicted in Fig. 3.2). Further,

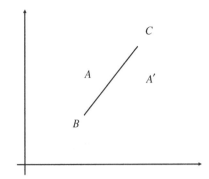

Fig. 3.2.   The expected utility of a risky gamble.

assume that points B and C represent a risky gamble with probabilities $0.50, 0.50$. The producer can tell whether he prefers the risk-free point A and the risky gamble B/C. To complete the proof, the cases where A is preferred to both B and C and B and C are preferred to A are ruled out. Mathematically, either A is preferred to the gamble or lottery between B and C ($A \succ B \circ C$) or the lottery is preferred to A ($B \circ C \succ A$). As a second postulate, as depicted in the previous example, a probability can be defined that makes the lottery indifferent with the certain payoff.

3.2.2.   *Conceptual Structure of the Axiomatic Treatment of Numerical Utilities*

In an axiomatic treatment given a set of axioms or basic notions are acceptable, the analyst shows that a conclusion follows using logical deductions based on these axioms or notions (see Debreu (1986) and Mirowski (1991) for a more complete history of mathematical rigor in economics). In this case, the economist wants to show that there exists a utility mapping $U(Y)$ such that if X is preferred to Z ($X \succ Z$) then $U(X) > U(Z)$ where X and Y are defined as risky lotteries.

Axioms of expected utility:

- $u \succ v$ is a complete ordering of U [3:A].
  - For any two u, v one and only one of the three following relations hold $u \approx v, u \succ v, u \prec v$ two outcomes [3:A:a].

o $u \succ v, v \succ w$ implies $u \succ w$. Basically, the axiom assumes that preferences are transitive [3:A:b].

- Ordering and combining [3:B]

  o $u \prec v$ implies $u \prec \alpha u + (1-\alpha)v$. This implies that if v is preferred to u, then a small amount of u combined with a large amount of v is preferred to u [3:B:a].
  o $u \succ v$ implies $u \succ \alpha u + (1-\alpha)v$ [3:B:b].
  o $u \prec w \prec v$ implies the existence of α such that $\alpha u + (1-\alpha)v \prec w$. This is the figure above, if w is "bounded" by u and v, then we can choose a α that makes the decision maker indifferent [3:B:c].
  o $u \succ w \succ v$ implies the existence of α such that $\alpha u + (1-\alpha)v \succ w$ [3:B:d]

- Algebra of combining [3:C]

  o $\alpha u + (1-\alpha)v = (1-\alpha)v + \alpha u$ [3:C:a].
  o $\alpha(\beta u + (1-\beta)v) + (1-\alpha)v = \gamma u + (1-\gamma)v$ where $\gamma = \alpha\beta$. This is an iterated gamble. The first part of the left-hand side is the utility of a risky payoff [3:C:b].

The terms in brackets denote von Neumann and Morgenstern's number of the axioms.

Appendix D gives a more detailed tour of the von Neumann and Morgenstern proof. However, a simplified version of the proof will be presented here. As a starting point, consider two lotteries. One lottery is an index lottery that defines the boundaries of possible outcomes $u_0(1-\alpha) + v_0\alpha$. The second lottery is the alternative that the decision maker wants to rank $w \approx (1-\beta)\tilde{u} + \beta\tilde{v}$. As a first step, determine if the alternative that the individual wants to rank is contained in the range of possible alternatives that can be ranked ($u_0 \prec \tilde{u} \prec \tilde{v} \prec v_0$). Next, given completeness, any risky alternative can be ranked against the index bundle (either $w \approx (1-\beta)\tilde{u} + \beta\tilde{v} \succ (1-\alpha)u_0 + \alpha v_0$ or $w \approx (1-\beta)\tilde{u} + \beta\tilde{v} \prec (1-\alpha)u_0 + \alpha v_0$). Given that each set is non-empty (axioms 3:B:c and 3:B:d above), α_0 can be defined that makes the decision maker indifferent between the index lottery and the risky alternative. This $\alpha_0 \in [0, 1]$ by definition of the lottery, thus it maps from preferences to real number space

$$\alpha_0 : (1 - \alpha_0)u_0 + \alpha_0 v_0 \approx (1 - \beta)\tilde{u} + \beta\tilde{v} \to [0, 1]. \qquad (3.28)$$

In addition, if the original lottery is recombined such that $\tilde{w} \approx (1 - \beta)\tilde{u} + \beta\tilde{v} \succ w$, then $\tilde{\alpha}_0 \to (1 - \tilde{\alpha}_0)u_0 + \tilde{\alpha}_0 v_0 \approx (1 - \tilde{\beta})\tilde{u} + \tilde{\beta}\tilde{v}$ is such that $\tilde{\alpha}_0 > \alpha_0$. In this case, α_0 is a measure of expected utility, which is the measure defined in Eq. (3.28) can be used to rank risky alternatives based on the index lottery. The expected utility provides a monotonically increasing measure of preference over the lotteries. The measure is made more general since any monotonically increasing function can be mapped into the range of the function. Thus, inverse mapping admits more general utility functions such as a power utility function depicted in Eq. (3.19).

3.2.3. *Analytical Examples of Expected Utility*

According to the von Neumann and Morgenstern proof, decision makers prefer actions that yield higher expected utility. Linking this with the results on utility maximization from microeconomic theory, preferences possess a variety of characteristics required for optimization. For example, utility is typically assumed to be positively monotonic and concave. One such set of preferences is the Cobb–Douglas utility function that yields an indirect utility function, which is roughly the same as the power utility function presented in Eq. (3.19). To complete a specification of decision making under risk, a probability density function (as developed in Chapter 2) is combined with a specific algebraic form of the utility function. Some combinations of utility and distribution functions have "*closed form*" or "*analytical solutions.*"

Several of these analytical solutions involve combinations of utility and distribution functions, which yield mean–variance solutions. In mean–variance solution, the decision maker makes choices among risky alternatives by comparing the ratio of the expected returns to the variance. Intuitively, alternatives yield higher expected utility if the expected return increases relative to variance.

One such closed form solution is derived by assuming that the utility function is quadratic. Taking the expected value of a *quadratic utility function* yields a function of the first two moments of the distribution

$$E[a + bY + cY^2] = \int_{-\infty}^{\infty} (a + bY + cY^2)f(Y)dY$$

$$= a\int_{-\infty}^{\infty} f(Y)dY + b\int_{-\infty}^{\infty} Yf(Y)dY$$

$$+ c\int_{-\infty}^{\infty} Y^2 f(Y)dY$$

$$= a + b\mu_Y + c(\mu_Y^2), \tag{3.29}$$

where μ_Y is the mean of the distribution and μ_Y^2 is the second non-central moment of Y. Substituting for the variance of the distribution

$$E[a + bY + cY^2] = a + b\mu_Y + c(\sigma_Y^2 + (\mu_Y)^2)$$

$$= a + b\mu_Y + c(\mu_Y)^2 + c\sigma_Y^2 \tag{3.30}$$

so that the expected utility is a function of the mean and the variance of the distribution. Typically it is assumed that $b > 0$ and $c < 0$ so that the utility function is increasing in income (at least to some point) and concave in income. Note that this result does not require the assumption that the returns are normally distributed.

Using a slightly different approach, the expected value of any utility function results in a mean–variance problem given that income is normally distributed. To demonstrate this point, consider discussion of the moment generating function of random variables in Chapter 2. The moment generating function for the normal distribution is defined as

$$M_X(t) = \exp\left(\mu t + \frac{1}{2}\sigma^2 t^2\right). \tag{3.31}$$

Since the normal distribution is completely defined by its first two moments, the expectation of any distribution function is a function of the mean and variance. Put slightly differently, if the returns to the investment opportunity are normally distributed, utility must be a function of the mean and the variance of the distribution because these parameters define the distribution function.

A more specific solution involves the use of the normal distribution function with the negative exponential utility function. Under these assumptions, the expected utility has a specific form that relates the expected utility to the mean and variance. Starting with the negative exponential utility function

$$U(x) = -\exp(-\rho x). \tag{3.32}$$

The expected utility can then be written as

$$E[U(x)] = \int_{-\infty}^{\infty} -\exp[-\rho x] f(x; \mu, \sigma^2) dx$$

$$= \int_{-\infty}^{\infty} -\exp[-\rho x] \frac{1}{\sigma\sqrt{2\pi}} \exp\left[-\frac{(x-\mu)^2}{\sigma^2}\right] dx. \tag{3.33}$$

Combining the exponential terms and taking the constants outside the integral yields:

$$E[U(x)] = -\frac{1}{\sigma\sqrt{2\pi}} \int_{-\infty}^{\infty} \exp\left[-\frac{1}{2}\left(\frac{x-\mu}{\sigma}\right)^2 - \rho x\right] dx. \tag{3.34}$$

Next substitute the a tranformation for the normalized deviation from the mean used to define a standard normal

$$z = \frac{x-\mu}{\sigma} \tag{3.35}$$

into Eq. (3.34). The inverse mapping for this transformation becomes

$$x = z\sigma - \mu \tag{3.36}$$

and the Jacobian is

$$dx = \sigma\, dz. \tag{3.37}$$

The transformed expectation in Eq. (3.34) can then be expressed as

$$z = \frac{x-\mu}{\sigma} \Rightarrow z\sigma = x - \mu \Rightarrow x = \mu + z\sigma$$

$$\therefore E[U(z)] = -\frac{1}{\sigma\sqrt{2\pi}} \int_{-\infty}^{\infty} \sigma \exp\left[-\frac{1}{2}z^2 - \rho z\sigma - \rho\mu\right] dz. \tag{3.38}$$

Next, complete the square

$$-\frac{1}{2}z^2 - \rho\sigma z - \rho\mu = \left\{-\frac{1}{2}z^2 - \rho\sigma z - \frac{1}{2}\rho^2\sigma^2\right\} + \frac{1}{2}\rho^2\sigma^2 - \rho\mu.$$

(3.39)

Substituting the implications of the completed square into Eq. (3.38) yields

$$E[U(z)] = -\frac{1}{\sqrt{2\pi}} \int_{-\infty}^{\infty} \exp\left[-\rho\left(\mu - \frac{\rho}{2}\sigma^2\right) - \frac{1}{2}(z - \rho\sigma)^2\right] dz$$

$$= -\exp\left(-\rho\left(\mu - \frac{\rho}{2}\sigma^2\right)\right) \frac{1}{\sqrt{2\pi}} \int_{-\infty}^{\infty} \exp\left(-\frac{1}{2}(z - \rho\sigma)^2\right) dz$$

$$= -\exp\left(-\rho\left(\mu - \frac{\rho}{2}\sigma^2\right)\right).$$

(3.40)

Thus, Eq. (3.40) gives an analytical solution of expected utility assuming that income is distributed normally. In this case, expected utility is determined as a function of the mean and the variance of the random variable. However, unlike the generic conjecture based on the moment generating function in Eq. (3.31), the closed-form solution of expected utility in terms of the mean, variance, and ρ parameters is relatively simple and straightforward.

To develop another closed form solution for expected utility, consider the power form of the utility function presented in Eq. (3.19), but this time assume that income is distributed log-normally. In this case, the expected utility is written as

$$E[U(\Upsilon)] = \frac{1}{\sigma\sqrt{2\pi}} \int_0^\infty \frac{\Upsilon^{1-r}}{1-r} \left(\frac{1}{\Upsilon}\exp\left[-\frac{1}{2\sigma^2}(\ln(\Upsilon) - \mu)^2\right]\right) d\Upsilon.$$

(3.41)

Transforming the regression back into X space where $X = \exp(\Upsilon)$ yields

$$E[U(X)] = \frac{1}{1-r}\frac{1}{\sigma\sqrt{2\pi}} \int_{-\infty}^\infty \exp((1-r)X)$$

$$\times \exp\left(-\frac{1}{2\sigma^2}(X - \mu)^2\right) dx$$

(3.42)

(note that $\Upsilon^{1-r} \Rightarrow \exp((1-r)\ln(\Upsilon))$). Thus, adding the terms in the exponential yields

$$E[U(X)] = \frac{1}{1-r}\frac{1}{\sigma\sqrt{2\pi}}$$

$$\times \int_{-\infty}^{\infty} \exp\left(\frac{-X^2 + 2X\mu - \mu^2 + 2(1-r)\sigma^2 X}{2\sigma^2}\right) dx$$

$$= \frac{1}{1-r}\frac{1}{\sigma\sqrt{2\pi}}$$

$$\times \int_{-\infty}^{\infty} \exp\left(\frac{-X^2 + 2X(\mu + (1-r)\sigma^2) - \mu^2}{2\sigma^2}\right) dx$$

$$E[U(X)] = \frac{1}{1-r}\frac{1}{\sigma\sqrt{2\pi}}$$

$$\times \int_{-\infty}^{\infty} \exp\left(-\frac{(X - (\mu + (1-r)\sigma^2))^2}{2\sigma^2}\right.$$

$$\left. + \mu(1-r) + (1-r)^2\frac{\sigma^2}{2}\right) dx$$

$$= \frac{1}{1-r}\exp\left[(1-r)\left(\mu + (1-r)\frac{\sigma^2}{2}\right)\right]. \qquad (3.43)$$

Note that if $r > 1$ the solution in Eq. (3.43) resembles the solution of the negative exponential in Eq. (3.39).

3.3. Expected Value — Variance and Expected Utility Models

Meyers (1987) demonstrates the consistency between the expected utility model and expected value–variance decision rules. The point of the paper is that neither quadratic utility nor normality is acceptable criteria from an empirical point of view for equating expected utility with expected value–variance models. The alternative developed by Meyers involves location and scale relationships within the distribution functions.

Definition 3.1. Two cumulative distribution functions $G_1(\cdot)$ and $G_2(\cdot)$ are said to differ only by location and scale parameters α and β if $G_1(x) = G_2(\alpha + \beta x)$ with $\beta > 0$.

Several two-parameter families satisfy this property such as the normal and the uniform distributions.

Given that the location and scale conditions can be used to guarantee that two distributions are the same, it also stands to reason that a similar rule can be used to guarantee that preference orderings are the same. Assume that there exists various choice sets, Υ_i, that only differ by location and scale parameters. Let X be the random variable of one of these alternatives derived by normalizing by the mean and the standard deviation of the alternative

$$X_i = \frac{\Upsilon_i - \mu_i}{\sigma_i}. \tag{3.44}$$

The expected utility of this alternative can then be derived as

$$EU[\Upsilon_i] = \int_a^b u(\mu_i + \sigma_i x) dF(x)$$

$$\equiv V(\sigma_i, \mu_i). \tag{3.45}$$

This expression transforms any utility function or distribution function into mean–variance preferences.

The first step in examining the properties of $V(\sigma, \mu)$ is to examine its partial derivatives. First, taking the derivative of Eq. (3.45) with respect to σ (applying Leibniz's rule and assuming that the bounds of integration do not change) yields

$$V_\sigma(\sigma, \mu) = \int_a^b u'(\mu + \sigma x) x \, dF(x)$$

$$= \int_a^b \left(u''(\mu + \sigma x) \int_a^x x \, dF(x) \right) dx. \tag{3.46}$$

Taking the derivative of Eq. (3.45) with respect to μ yields

$$V_\mu(\sigma, \mu) = \int_a^b u'(\mu + \sigma x) dF(x). \tag{3.47}$$

Property 1: $V_\mu(\sigma, \mu) \geq 0$ for all μ and all $\sigma \geq 0$ if and only if $u'(\mu + \sigma x) \geq 0$ for all $\mu + \sigma x$.

Property 2: $V_\sigma(\sigma, \mu) \leq 0$ for all μ and all $\sigma \geq 0$ if and only if $u''(\mu + \sigma x) \leq 0$ for all $\mu + \sigma x$.

Property 1 states that increases in μ increase $V(\,\cdot\,)$ if the marginal utility $u'(\,\cdot\,)$ is positive. Similarly, increases in σ reduce $V(\,\cdot\,)$ if the utility function is concave. Totally differentiating $V(\,\cdot\,)$ yields

$$dV(\sigma, \mu) = V_\sigma(\sigma, \mu)d\sigma + V_\mu(\sigma, \mu)d\mu = 0$$

$$V_\mu(\sigma, \mu)d\mu = -V_\sigma(\sigma, \mu) \tag{3.48}$$

$$\frac{d\mu}{d\sigma} = -\frac{V_\sigma(\sigma, \mu)}{V_\mu(\sigma, \mu)} = S(\sigma, \mu),$$

where $S(\sigma, \mu)$ is the relationship (isoquant) between the location parameter (μ) and scale (σ) that leaves the decision maker utility unchanged. Note that $S(\sigma, \mu)$ is

$$S(\sigma, \mu) = \frac{-V_\sigma(\sigma, \mu)}{V_\mu(\sigma, \mu)}$$

$$= \frac{-\int_a^b \left(u''(\mu + \sigma x) \int_a^x x \, dF(x) \right) dx}{\int_b^a u'(\sigma + \mu x) dF(x)}. \tag{3.49}$$

Property 3: $S(\sigma, \mu) \geq 0$ for all μ and all $\sigma \geq 0$ if and only if $u'(\mu + \sigma x) \geq 0$ and $u''(\mu + \sigma x) \leq 0$ for all $\mu + \sigma x$.

Since $V(\cdot)$ is an ordinal representation of preferences, it is necessary to demonstrate that $V(\,\cdot\,)$ is concave. This result is established by

$$V_{\mu\mu}(\sigma, \mu) = \int_a^b u''(\mu + \sigma x) dF(x)$$

$$V_{\mu\sigma}(\sigma, \mu) = \int_a^b u''(\mu + \sigma x)x \, dF(x) \tag{3.50}$$

$$V_{\sigma\sigma}(\sigma, \mu) = \int_a^b u''(\mu + \sigma x)x^2 \, dF(x).$$

Property 4: $V(\sigma, \mu)$ is a concave function for all μ and all $\sigma \geq 0$ if and only if $u''(\mu + \sigma x) \leq 0$ for all $\mu + \sigma x$.

Concavity is established by the fact that both $V_{\mu\mu}(\cdot)$ and $V_{\sigma\sigma}(\cdot)$ are both nonpositive if $u''(\cdot)$ is negative and $V_{\mu\mu}(\cdot)V_{\sigma\sigma}(\cdot) - V_{\mu\sigma}^2(\cdot)$ is nonnegative.

Property 5: $\partial S(\sigma, \mu)/\partial \mu \leq (\ =, \geq \ )0$ for μ and all $\sigma > 0$ if and only if $u(\mu + \sigma x)$ displays decreasing (constant, increasing) absolute risk aversion for all $\mu + \sigma x$.

Property 6: $dS(t\sigma, t\mu)/dt < (\ =, > \ )0$ for μ and all $\sigma > 0$ if and only if $u(\mu + \sigma x)$ displays decreasing (constant, increasing) absolute risk aversion for all $\mu + \sigma x$.

This property indicates that the slope of the indifference curve changes as you scale the distance from the origin depending on the individual's relative risk aversion.

Property 7: $S_1(\sigma, \mu) \geq S_2(\sigma, \mu)$ for all (σ, μ) if and only if $u_1(\mu + \sigma x)$ is more risk averse than $u_2(\mu + \sigma x)$.

3.4. Problems with Expected Utility

The problems with the expected utility approach are primarily based on empirical anomalies. As described by Machina (1987), these difficulties are traced to Allais' Paradox, which raises difficulties with the independence axiom of the von Neumann–Morgenstern approach. To develop this paradox, this section starts by developing Machina's formulation of expected utility.

Machina demonstrates how the expected utility formulation is linear in probabilities. Start by assuming that there are three possible outcomes for a random variable $\{x_1, x_2, x_3\}$ with $x_1 < x_2 < x_3$ and probabilities $P = (p_1, p_2, p_3)$. Given this formulation

$$\sum_{i=1}^{3} p_i = 1 \Rightarrow p_2 = 1 - p_1 - p_2, \tag{3.51}$$

which implies that

$$U(x) = p_1 U(x_1) + p_2 U(x_2) + p_3 U(x_3)$$
$$= p_1 U(x_1) + (1 - p_1 - p_3) U(x_2) + p_3 U(x_3). \qquad (3.52)$$

Holding the payoffs in Eq. (3.52) constant and differentiating with respect to the levels of probability yields

$$dU(x) = U(x_1) dp_1 + (-dp_1 - dp_3) U(x_2)$$
$$+ U(x_3) dp_3 = 0$$
$$[U(x_1) - U(x_2)] dp_1 = -[-U(x_2) + U(x_3)] dp_3 \qquad (3.53)$$
$$\frac{[U(x_2) - U(x_1)]}{[U(x_3) - U(x_2)]} = \frac{dp_3}{dp_1}$$

which given that $x_1 < x_2 < x_3$ and if $U'(x) > 0$ implies that any increase in p_3 relative to p_1 that holds the total probability constant (i.e., the probabilities still sum to one) yields more expected utility. Completing the model, construct a mean-preserving spread (i.e., change the probability in such a way that leaves the mean of the risky alternative unchanged, but changes the variance).

$$\bar{x} = \sum_{i=1}^{3} x_i p_i = x_1 p_1 + x_2 p_2 + x_3 p_3$$
$$= x_1 p_1 + x_2 (1 - p_1 - p_3) + x_3 p_3 \qquad (3.54)$$
$$d\bar{x} = x_1 dp_1 + x_2 (-dp_1 - dp_3) + x_3 dp_3 = 0$$
$$\frac{[x_2 - x_1]}{[x_3 - x_2]} = \frac{dp_3}{dp_2}.$$

Figure 3.3 presents both of these relationships mapped onto the two-dimensional space spanned by p_3 and p_1. In this graph, any point to the northwest is preferred to any point to the southeast. Thus, the solid lines represent relationships between the probabilities that leave the expected value of the risky alternative unchanged. The dashed lines (the steeper lines) imply that the utility function is more concave so

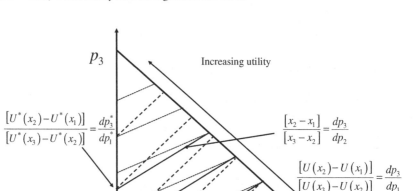

Fig. 3.3.   Linearity of probabilities and expected utility.

that the decision maker is relatively more risk averse (prefers relatively less risk) and the dotted lines (the flatter lines) imply that the utility function is less concave so that the decision maker is relatively less risk averse (prefers relatively less risk). Note that the steeper the iso-utility line (the more risk averse) the lower the optimum selection of p_3 and, hence, the less risky the choice of probabilities. On the other hand, the flatter the iso-utility line (the less risk averse) the greater the selection of p_3 and the more risky the overall choice of probabilities.

The independence assumption places restrictions on individual's preferences over mixtures of lotteries. Taking two sets of probabilities $P^* = (p_1^*, p_2^*, \ldots, p_n^*)$ and $P = (p_1, p_2, \ldots, p_n)$ two lotteries defined on a common event space $\{x_1, x_2, \ldots, x_n\}$, the mixture lottery is defined as

$$\alpha P^* + (1 - \alpha)P = (\alpha p_1^* + (1 - \alpha)p_1, \alpha p_2^*$$
$$+ (1 - \alpha)p_2, \ldots, \alpha p_n^* + (1 - \alpha)p_n). \quad (3.55)$$

This is a compound lottery that offers an α chance of winning a lottery with probabilities P^* and a $1 - \alpha$ chance of winning a lottery with probabilities P. Under linearity in probabilities

$$\sum_{i=1}^{n} U(x_i)(\alpha p_i^* + (1-\alpha)p_i) = \alpha \sum_{i=1}^{n} U(x_i)p_i^* + (1-\alpha) \sum_{i=1}^{n} U(x_i)p_i.$$

(3.56)

Given this structure, the Independence Axiom as described by Samuelson is

> If the lottery P^* is preferred (indifferent) to the lottery P, then the mixture $\alpha P^* + (1-\alpha)P^{**}$ will be preferred (indifferent) to the mixture $\alpha P + (1-\alpha)P^{**}$ for all $\alpha > 0$ and P^{**}.

This structure allows for the development of the systematic violation known as Allais' Paradox. Consider the two pairs of gambles

$$a_1 = \{1.00 \text{ chance of } \$1,000,000 \quad \text{versus}$$

$$a_2 = \begin{cases} 0.10 \text{ chance of } \$5,000,000 \\ 0.89 \text{ chance of } \$1,000,000 \\ 0.01 \text{ chance of } \$0 \end{cases}$$

(3.57)

and

$$a_3 = \begin{cases} 0.10 \text{ chance of } \$5,000,000 \\ 0.90 \text{ chance of } \$0 \end{cases} \quad \text{versus}$$

$$a_4 = \begin{cases} 0.11 \text{ chance of } \$1,000,000 \\ 0.89 \text{ chance of } \$0 \end{cases}.$$

(3.58)

Defining $\{x_1, x_2, x_3\} = \{\$0, \$1,000,000, \$5,000,000\}$, these four gambles can be depicted as a parallelogram within the triangular diagram as depicted in Fig. 3.4. In this case, the selection of a_1 implies that a_4 is preferred to a_3. However, numerous empirical studies have found a preference for a_1 and a_3. Such preferences require that the indifference curves "fan-out" so that the iso-utility curve becomes steeper as it approaches the p_3 axis and flatter as the iso-utility curve approaches the p_1 axis as depicted in the green lines in Fig. 3.4.

The Allais Paradox is a special case of a general empirical pattern called the *common consequence* effect. The common consequence effect

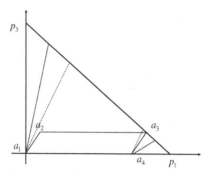

Fig. 3.4.   The Allais Paradox.

involves pairs of probability mixtures of the form

$$b_1 : \alpha \delta_x + (1 - \alpha) P^{**} \quad \text{versus} \quad b_2 : \alpha P + (1 - \alpha) P^{**} \tag{3.59}$$

and

$$b_3 : \alpha \delta_x + (1 - \alpha) P^* \quad \text{versus} \quad b_4 : \alpha P + (1 - \alpha) P^*, \tag{3.60}$$

where δ_x is a certain payoff, P is a lottery that pays more or less than δ_x, and P^{**} dominates P^*. In the above example, a_1 is b_1, a_2 is b_2, a_3 is b_4, and a_4 is b_3. In this example

$$P = \begin{bmatrix} \dfrac{10}{11} & \dfrac{1}{11} \end{bmatrix}, \quad Q = \begin{bmatrix} \$5{,}000{,}000 \\ \$0 \end{bmatrix}$$

$$\Rightarrow PQ = \$5{,}000{,}000 \dfrac{10}{11} + \$0 \dfrac{1}{11}$$

$$P^* = \begin{bmatrix} \dfrac{11}{11} & 0 \end{bmatrix}, \quad Q^* = \begin{bmatrix} \$0 \\ K \end{bmatrix}$$

$$\Rightarrow P^* Q^* = \$0 \dfrac{11}{11} + K \dfrac{0}{11}. \tag{3.61}$$

$$P^{**} = \begin{bmatrix} \dfrac{11}{11} & 0 \end{bmatrix}, \quad Q^{**} = \begin{bmatrix} \$1{,}000{,}000 \\ K \end{bmatrix}$$

$$\Rightarrow P^{**} Q^{**} = \$1{,}000{,}000 \dfrac{11}{11} + K \dfrac{0}{11}.$$

Building the gambles

$$b_1: \$1,000,000\alpha + (1-\alpha)\frac{11}{11}\$1,000,000 = \$1,000,000$$

$$b_2: \alpha \begin{bmatrix} \dfrac{10}{11} & \dfrac{1}{11} \end{bmatrix} \begin{bmatrix} \$5,000,000 \\ \$0 \end{bmatrix}$$

$$+ (1-\alpha) \begin{bmatrix} \dfrac{11}{11} & \dfrac{0}{11} \end{bmatrix} \begin{bmatrix} \$1,000,000 \\ K \end{bmatrix}$$

$$= \alpha \left(\frac{10}{11}\$5,000,000 + \frac{1}{11}\$0 \right)$$

$$+ (1-\alpha) \left(\frac{11}{11}\$1,000,000 + \frac{0}{11}K \right)$$

$$= \alpha\frac{10}{11}\$5,000,000 + (1-\alpha)\$1,000,000$$

$$b_3: \alpha\$1,000,000 + (1-\alpha) \begin{bmatrix} \dfrac{11}{11} & \dfrac{0}{11} \end{bmatrix} \begin{bmatrix} \$0 \\ K \end{bmatrix}$$

$$= \alpha\$1,000,000 + (1-\alpha) \left(\frac{11}{11}\$0 + \frac{0}{11}K \right)$$

$$= \alpha\$1,000,000 + (1-\alpha)\$0$$

$$b_4: \alpha \begin{bmatrix} \dfrac{10}{11} & \dfrac{1}{11} \end{bmatrix} \begin{bmatrix} \$5,000,000 \\ \$0 \end{bmatrix}$$

$$+ (1-\alpha) \begin{bmatrix} \dfrac{11}{11} & \dfrac{0}{11} \end{bmatrix} \begin{bmatrix} \$1,000,000 \\ K \end{bmatrix}$$

$$= \alpha \left(\frac{10}{11}\$5,000,000 + \frac{1}{11}\$0 \right)$$

$$+ (1-\alpha) \left(\frac{11}{11}\$1,000,000 + \frac{0}{11}K \right)$$

$$= \alpha\frac{10}{11}\$5,000,000 + \alpha\frac{1}{11}\$0 + (1-\alpha)\$1,000,000. \qquad (3.62)$$

Letting $\alpha = \frac{11}{100}$ the yields the gambles in Eqs. (3.57) and (3.58).

The salient question is then whether the fact that decision makers choose risky scenarios that are not consistent with the expected utility function implies that the von Neumann–Morgenstern proof is flawed. Note that the empirical failures occur at the extremes, potentially in regions in the risk/return space where the decision maker may not be able to make adequate comparisons. This would indicate a weakness in transitivity in the most extreme comparisons. While not attempting to sweep the debate under the rug, the remaining portions of this book accept (either explicitly or implicitly) the von Neumann–Morgenstern proof.

3.5.   Summary

This chapter develops the expected utility hypothesis, which is the theoretical basis for several important models of decision making under uncertainty. In addition, the chapter demonstrates how expected value–variance orderings are consistent with expected utility under a variety of assumptions including normality and location scale transformations. Further, the chapter presents the derivation of the combinations of normality and specific preference functions that allow for closed form solutions of the general expected utility model. Finally, the chapter briefly presents some of the objections to the expected utility framework.

Chapter 4

Risk Aversion in the Large and Small

The preceding chapter presented a fairly rigorous development of the expected utility hypothesis using the von Neumann–Morgenstern approach. Under the expected utility hypothesis, consumers will choose the alternative that produces the larger expected utility. However, such a conclusion admits a variety of optimizing behaviors. For example, the decision maker may be indifferent to risk, simply choosing the alternative that maximizes expected income. Such preferences are referred to as *risk neutral*. Alternatively, individuals may be averse to risk so that they are willing to pay to avoid risk. These preferences are termed *risk aversion* and are directly observable when individuals purchase insurance for protection against risk such as fire, automobile collision, and life insurance. Finally, it may be that economic agents prefer risk or are willing to pay to be exposed to risk. This behavior is referred to as risk-seeking or *risk-preferring* behavior. The frequency of such preferences is apparent in the proliferation of gambling in the United States in the 1990s. Further complicating, the discussion of risk aversion is the fact that a particular individual may exhibit each attitude toward risk, buying a lottery ticket, making any investment with a positive net present value, and insuring property against acts of nature. The analysis of risk this text emphasizes is decision under risk aversion limiting on risk neutrality. Thus, the discussion will focus on individuals who are willing to pay to avoid risk.

As a starting point of the discussion of risk aversion, consider the power utility function presented in Chapter 3

$$U[\Upsilon] = \frac{\Upsilon^{1-r}}{1-r}, \tag{4.1}$$

where Υ is the level of income and r, as will be seen in this discussion, is a parameter that captures the degree of risk aversion. Using the power utility function, compute the value r that yields a fixed risk premium. Again, relying on the discussion in Chapter 3, define the risk premium $(\pi[r])$ as

$$\pi(r) = E[\Upsilon] - CE(r), \tag{4.2}$$

where $E[\Upsilon]$ is the expected level of income and $CE(r)$ is the certainty equivalent conditional on the value of r. Next, define the certainty equivalent based on the expected utility. Assuming a Bernoulli distribution

$$E[U|r] = p\frac{\Upsilon_1^{1-r}}{1-r} + (1-p)\frac{\Upsilon_2^{1-r}}{1-r}, \tag{4.3}$$

where Υ_1 and Υ_2 are two random payoffs from this alternative and p is the probability that payoff Υ_1 will occur. Thus, the certainty equivalent becomes

$$CE(r) = \left[\left\{ p\frac{\Upsilon_1^{1-r}}{1-r} + (1-p)\frac{\Upsilon_2^{1-r}}{1-r} \right\} (1-r) \right]^{\frac{1}{1-r}}$$

$$= (p\Upsilon_1^{1-r} + (1-p)\Upsilon_2^{1-r})^{\frac{1}{1-r}}. \tag{4.4}$$

Putting the two parts together

$$\pi(r) = p\Upsilon_1 + (1-p)\Upsilon_2 - [p\Upsilon_1^{1-r} + (1-p)\Upsilon_2^{1-r}]^{\frac{1}{1-r}},$$

or

$$CE(r) = [p\Upsilon_1^{1-r} + (1-p)\Upsilon_2^{1-r}]^{\frac{1}{1-r}} = p\Upsilon_1 + (1-p)\Upsilon_2 - \pi(r). \tag{4.5}$$

Table 4.1.   Certainty equivalents, risk premia, and risk aversions for $100,000 and $50,000 gamble with equal probability of each.

Certainty equivalent	Risk premium	r
75,000	0	0.0000
72,500	2,500	0.5824
70,000	5,000	1.1683
69,500	5,500	1.2880
69,000	6,000	1.4092
68,500	6,500	1.5319
68,000	7,000	1.6565

Thus, fixing the payoffs of the gamble (Y_1, Y_2, and p), solve for the value of r that yields a given risk premium or certainty equivalent. Table 4.1 gives the certainty equivalent, risk premium, and r for a gamble that pays $100,000 with probability 0.5 and $50,000 with a probability of 0.5.

The relationship between the risk aversion, the risk premium, and certainty equivalent presented in Table 4.1 is determined by the relative concavity of the utility function. By varying r in Eq. (4.1), functions that are convex, linear, and concave can be generated. Specifically, taking the second derivative of the power utility function yields

$$\frac{\partial^2 U(Y)}{\partial Y^2} = -rY^{-(1+r)}. \qquad (4.6)$$

Thus, if $r < 0$, the second derivative of the power utility function is positive; the utility function is convex in income. The convexity of the utility function implies *risk-taking* behavior (or a negative risk premium). For example, letting $r = -0.2372$ yields a risk premium of −$1,000 (Fig. 4.1). If $r = 0$, the power utility function is linear in income, so the decision maker is *risk neutral.* Alternatively, for $r > 0$ the second derivative is negative, implying that the utility function is concave in income. Figure 4.2 depicts the expected utility, certainty equivalent, and risk premium of the same gamble with $r = 0.2341$. The results presented in Fig. 4.2 indicate a positive risk premium or

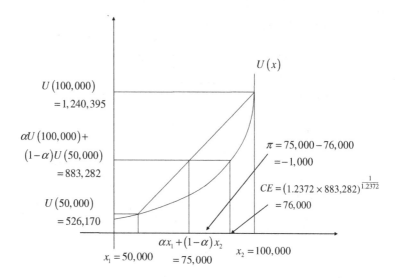

Fig. 4.1. Expected utility and certainty equivalent under risk seeking behavior.

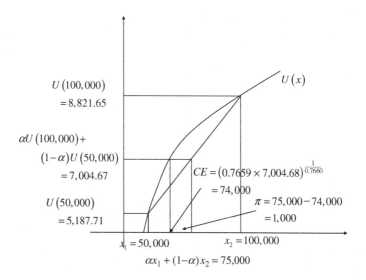

Fig. 4.2. Expected utility and certainty equivalent under risk aversion.

aversion to risk. Thus, risk aversion is intimately linked to the concavity of the utility function.

Theorem 4.1. *An agent is:*

Risk averse if and only if his von Neumann–Morgenstern utility function $U(\Upsilon)$ is concave,

Risk neutral if and only if his von Neumann–Morgenstern utility function $U(\Upsilon)$ is linear, and

An agent is strictly risk averse if and only if his von Neumann–Morgenstern utility function $U(\Upsilon)$ is strictly convex.

To justify this theorem, start with the basic concept of concavity. If the utility function is concave

$$U(\alpha x_1 + (1 - \alpha)x_2) \geq \alpha U(x_1) + (1 - \alpha)U(x_2). \qquad (4.7)$$

The implications of Eq. (4.7) are presented graphically in Fig. 4.3. Interpreting α as a probability completes the linkage between concavity and risk aversion.

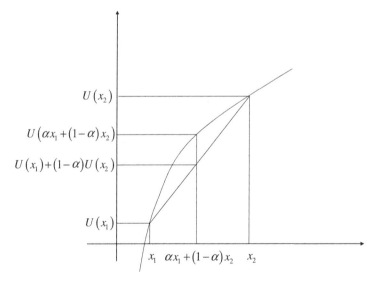

Fig. 4.3. Implications of concavity for risk aversion.

4.1.   Arrow–Pratt Risk Aversion Coefficient

To formalize our discussion of risk aversion, this section follows the approach of Pratt (1964). The risk premium $\pi(x, y)$ is the certain payment that leaves the decision maker indifferent between the gamble and the risk-free outcome. Mathematically, the risk premium is defined as

$$U(x + E(z) - \pi(x, z)) = E\{U(x + z)\}. \tag{4.8}$$

The first step is to normalize out the $E(z)$. We note that

$$\pi_a(x, z) = \pi(x + \mu, z - \mu). \tag{4.9}$$

Thus, the expected value of the risky gamble is just added to the base endowment x.

Note that the investor is indifferent between receiving the risky return z and the sure amount

$$\pi_a(x, z) = E(z) - \pi(x, z) \tag{4.10}$$

$\pi_a(x, z)$ is called the certainty equivalent of the gamble (sound familiar?). The certainty equivalent is defined by

$$U(x + \pi_a(x, z)) = E\{U(x + z)\}. \tag{4.11}$$

4.1.1.   *Local Risk Aversion*

Given this formulation, it is possible to define the local risk aversion for an investor. Going back to the original equation for the risk premium, ignoring for the moment $E(z)$, the utility function can be written as

$$U(x - \pi) = U(x) - \pi U'(x) + O(\pi^2). \tag{4.12}$$

The right-hand side of this expression is simply the first-order Taylor expansion. Taking the second-order expansion of the expected utility on the right-hand side of Eq. (4.7) yields

$$E\{U(x + z)\} = E\left\{U(x) + zU'(x) + \frac{1}{2}z^2 U''(x) + O(z^3)\right\}$$

$$= U(x) + \frac{1}{2}\sigma_z^2 U''(x) + \alpha(\sigma_z^2). \tag{4.13}$$

Setting these expressions equal to one another, as depicted in the equation yields

$$U(x) - \pi U'(x) = U(x) + \frac{1}{2}\sigma_z^2 U''(x)$$

$$\pi(x, z) = \frac{1}{2}\sigma_z^2 \frac{U''(x)}{U'(x)}. \qquad (4.14)$$

The last part of the expression defines local risk aversion

$$r(x) = -\frac{U''(x)}{U'(x)} = -\frac{d}{dx}\ln(U'(x)). \qquad (4.15)$$

From this formulation, it is apparent that the primary measure of the relative risk aversion involves the concavity of the utility function. There are two dominant measures, the *absolute Arrow–Pratt risk aversion coefficient*

$$R_A(x) = -\frac{U''(x)}{U'(x)} \qquad (4.16)$$

and the *relative Arrow–Pratt risk aversion coefficient*

$$R_R(x) = -\frac{xU''(x)}{U'(x)}. \qquad (4.17)$$

Two utility functions demonstrate the alternative forms of risk aversion. The first utility function is the negative exponential utility function that depicts constant relative risk aversion. Specifically,

$$U(x) = -\exp(-\rho x). \qquad (4.18)$$

The absolute risk aversion coefficient becomes

$$R_A(x) = -\frac{-\rho^2 \exp(-\rho x)}{\rho \exp(-\rho x)} = \rho. \qquad (4.19)$$

The utility function that possesses constant relative risk aversion is the power utility function depicted as

$$U(x) = \frac{x^{1-r}}{1-r}. \qquad (4.20)$$

Table 4.2.   Expected value and certainty equivalence for hypothetical gambles.

Initial equity	Expected value	$r = 0.25$	$r = 0.01$	$\theta = 0.00025$
$10,000	$11,625	$11,543	$6,574	$11,300
$100,000	$116,250	$115,432	$65,743	$95,543

The constant relative risk aversion coefficient then becomes

$$R_R(x) = -x \frac{-rx^{-r-1}}{x^{-r}} = r. \tag{4.21}$$

The difference between the two can be demonstrated by looking at the value of two representative gambles. The first gamble is based on a wealth of $10,000. There are two possible outcomes for this gamble. The first outcome yields 90% of the decision maker's original wealth or $9,000. The second potential outcome yields 1.25% of the decision maker's original wealth or $12,500. Next, consider a second gamble based on the same probabilities and rates of return, but assume that the second individual is endowed with $100,000. Thus, this gamble pays $90,000 in the first outcome and $125,000 in the second outcome.

Table 4.2 gives the expected value and certainty equivalence for each scenario under constant relative risk aversion. These scenarios show the proportionality between the constant relative risk aversion scenarios while no such proportionality is demonstrated under the constant absolute risk aversion scenario.

Another manifestation of the different implications of these functions is the absolute risk aversion coefficient of the power utility function. Specifically,

$$R_A(x) = -\frac{rx^{-r-1}}{x^{-r}} = \frac{r}{x}. \tag{4.22}$$

4.1.2. *Transformations of Scale for the Arrow–Pratt Absolute Risk Aversion Coefficient*

The forgoing analysis demonstrates how risk aversion can be encapsulated into a single parameter. The question then becomes: where do analysts obtain these coefficients for empirical work? Two sources of information about risk aversion are published literature or by directly estimating risk aversion using either primary or secondary data. This section focuses on the use of published coefficients. Specifically, the use of risk aversion coefficients estimated or elicited for one analysis may not be appropriate for use in another application without attention to scaling of the risky alternative. Raskin and Cochran (1986) address many of these scaling issues.

Returning to the basic question, under the expected utility hypothesis, alternative a_1 is preferred to alternative a_2 if

$$E[U(x|a_1)] > E[U(x|a_2)]. \tag{4.23}$$

Assuming a negative exponential utility function and normally distributed returns for each alternative, this decision is equivalent to

$$\mu_1(x) - \frac{\rho}{2}\sigma_1^2(x) > \mu_2 - \frac{\rho}{2}\sigma_2^2(x), \tag{4.24}$$

where $\mu_1(x)$ and $\mu_2(x)$ are the expected returns for alternative 1 and alternative 2, respectively, $\sigma_1^2(x)$ and $\sigma_2^2(x)$ are the variances for each alternative, and ρ is the Arrow–Pratt absolute risk aversion coefficient. Under these simplifying assumptions, the risk aversion level where the decision maker prefers a_1 to alternative a_2 can be derived as

$$[\mu_1(x) - \mu_2(x)] > -\frac{\rho}{2}[\sigma_2^2(x) - \sigma_1^2(x)]$$
$$\rho > \frac{2[\mu_1(x) - \mu_2(x)]}{[\sigma_2^2(x) - \sigma_1^2(x)]}. \tag{4.25}$$

Thus, the preference relationship is linked directly to the risk aversion coefficient. This linkage is intimately related to the scaling of the risk aversion coefficient. First, assume that the Arrow–Pratt absolute risk aversion coefficient taken from the literature is defined as

$$\rho_z = R_A(z) = -\frac{U''(z)}{U'(z)} \tag{4.26}$$

for some gamble defined around income z. For example, suppose that the Arrow–Pratt absolute risk aversion coefficient is defined by a tangency on the expected value–variance frontier (to be further developed in Chapter 5)

$$\rho = \frac{1}{2}\frac{\mu(z)}{\sigma^2(z)} \tag{4.27}$$

where z is defined as a rate of return.

Returning to the basic utility maximization problem in Eq. (4.23), recognize that the scale of the payoff may be different than the scale used to define the risk aversion in Eq. (4.27). Specifically, while the risk aversion coefficient was defined based on the rate of return, the desired scale in Eq. (4.23) could be the absolute risk aversion coefficient for farm income (i.e., in dollars per year). One alternative would be to rescale the return in Eq. (4.27) to be representative of the gamble in Eq. (4.23), or

$$x = a + bz, \tag{4.28}$$

where a could be a constant source of farm income and b the level of wealth invested in the farm enterprise. To relate the risk aversion coefficient for the rate of return specification, solve for the value of z as a function of the farm income specification

$$x = a + bz \Rightarrow z = \frac{x - a}{b}. \tag{4.29}$$

Thus, substituting this transformation into the original (i.e., z valued) utility function

$$U\left(\frac{x-a}{b}\right) \Rightarrow \frac{\partial U \frac{(x-a)}{b}}{\partial x} = \frac{1}{b}U'\left(\frac{x-a}{b}\right) = \frac{1}{b}U'(z)$$

$$\Rightarrow \frac{\partial^2 U \frac{(x-a)}{b}}{\partial x^2} = \frac{1}{b^2}U''\left(\frac{x-a}{b}\right) = \frac{1}{b^2}U''(z). \tag{4.30}$$

Based on these derivatives

$$\rho_x = -\frac{\frac{1}{b^2}U''(z)}{\frac{1}{b}U'(z)} = -\frac{1}{b}\frac{U''(z)}{U'(z)} = \frac{1}{b}\rho_z \tag{4.31}$$

or alternatively,

$$\rho_z = b\rho_x. \tag{4.32}$$

Note that the scaling of the outcome range should not be confused with an arbitrary positive linear transformation of the utility function. Specifically, both the selection of the optimal alternative and the relative risk aversion coefficient are invariant to positive linear transformation. To see this let

$$V(x) = a + bU(x), \tag{4.33}$$

where $V(x)$ is a new utility function formed by a linear transformation of the original utility function, $U(x)$, such that a is any constant and b is any positive constant. Given this transformation

$$U(x) > U(y) \Rightarrow a + bU(x) > a + bU(y)$$
$$\Rightarrow V(x) > V(y), \tag{4.34}$$

so the utility maximizing choice is invariant to a positive linear transformation. Further,

$$
\begin{aligned}
R_A(x) &= -\frac{V''(x)}{V'(x)} \\
&= -\frac{\frac{\partial^2 (a+bU(x))}{\partial x^2}}{\frac{\partial (a+bU(x))}{\partial x}} \\
&= -\frac{bU''(x)}{bU'(x)} = -\frac{U''(x)}{U'(x)} \tag{4.35}
\end{aligned}
$$

so that the risk aversion coefficient is also invariant to a positive linear transformation of the utility function.

Raskin and Cochran begin their discussion of the effect of transformations of the risk aversion coefficient by developing an interpretation of the Pratt–Arrow coefficient in terms of marginal utility

$$
\begin{aligned}
r(x) &= -\frac{u''(x)}{u'(x)} \\
&= -\frac{\left(\frac{du'(x)}{dx}\right)}{u'(x)}
\end{aligned}
$$

$$= -\frac{d}{dx}\ln(u'(x))$$

$$= -\frac{\left(\frac{du'(x)}{u'(x)}\right)}{dx}. \tag{4.36}$$

Thus, the absolute risk aversion coefficient is the percent change in the marginal utility level for a given level of income. Therefore, r is associated with a unit of change in outcome space. If the risk aversion coefficient was elicited in outcomes of dollars, then the risk aversion coefficient is 0.0001/$. This result indicates that the decision maker's marginal utility is falling at a rate of 0.01% per dollar change in income. This association between the risk aversion and the level of income then raises the question of the change in outcome scale.

Given that the absolute risk aversion coefficient is determined by the level of income, the scaling of risk aversion coefficients based on the gamble can be formalized from Theorem 4.1.

Theorem 4.2. *Let $R_A(x) = U''(x)/U'(x)$. Define a transformation of scale on x such that $z = x/b$, where b is a constant. Then $R_A(z) = bR_A(x)$.*

The proof lies in the change in variables. Given

$$U(x) = U(bz)$$

$$\frac{dU}{dz} = \frac{dU}{dx}\frac{dx}{dz} = bU'(z) \tag{4.37}$$

$$\frac{d^2U}{dz^2} = \frac{d}{dz}\left(\frac{dU}{dz}\right) = \frac{d}{dz}(bU'(z)) = b^2 U''(x).$$

In other words, if the scale of the outcome changes by b, the scale of the risk aversion coefficient must be changed by the same amount.

Theorem 4.3. *If $z = x + a$, where a is a constant, then $R_A(z) = R_A(x)$. Therefore, the magnitude of the risk aversion coefficient is unaffected by the use of incremental rather than absolute returns.*

Example: Suppose that a study of US farmers gives a risk aversion coefficient of $r = 0.0001/$ (the United States). Application to the

Australian farmers whose dollar is worth 0.667 of the US dollar is $r = 0.0000667/\$$ Australian.

4.2. Eliciting Risk Aversion Coefficients

One alternative to the use of published risk aversion coefficients is to elicit risk aversion coefficients directly from the group of decision makers or specific decision maker of interest. It must be recognized, however, that this elicitation can be a time consuming and costly process.

4.2.1. *Direct Elicitation of Utility Functions*

While the direct elicitation of utility can be used with single-valued utility functions, lexicographic utility functions, or broader concepts of multiple goals, it is typically applied using variations of von Neumann and Morgenstern's expected utility model. Building on the standard expected utility formulation, the direct elicitation method asks the decision maker whether they prefer one gamble (implying one risk aversion coefficient) to another (implying another risk aversion coefficient). Based on that choice, another gamble is then presented to the decision maker with a systematic change in the implied risk aversion. The process continues until the decision maker's risk aversion coefficient is determined to a desired degree of precision (i.e., bounded within a specific tolerance).

In this formulation, Y_1 and Y_2 are possible random outcomes with the probability of Y_1 being p and the probability of Y_2 being $1 - p$. Based on this probability, the decision maker is asked whether they prefer the risky alternative to a certain payoff of $\tilde{Y}$. Going back to the previous examples, suppose that the decision maker is given the alternative of a 50/50 bet of paying either $50 or $5 (with an expected return of $27.50) versus a certain payoff of $26. If the answer is yes (i.e., they prefer the risky alternative to the certain outcome), the decision maker's risk aversion coefficient is less than 0.005944.

$$
\begin{aligned}
E[U(Y)] &= -\exp(-\rho Y_1)(p) - \exp(-\rho Y_2)(1 - p) \\
&= -\exp(-\rho \tilde{Y}).
\end{aligned} \tag{4.38}
$$

This system can be solved numerically using Gauss–Seidel

$$\exp(-\rho\tilde{Y}) - \exp(-\rho Y_1)(p) - \exp(-\rho Y_2)(1-p) = 0$$
$$\exp(-\rho 26.0) - 0.50\exp(-\rho 5) - 0.50\exp(-\rho 50) = 0. \tag{4.39}$$

If the answer is no (i.e., they prefer the certain outcome to the risky alternative), the researcher can take a couple of different approaches. First, the analyst could change the probabilities by making the higher return more likely (i.e., let $p = 0.47$). If the decision maker is willing to take that gamble, a new risk aversion coefficient could be derived as

$$\exp(-\rho 26.0) - 0.40\exp(-\rho 5) - 0.60\exp(-\rho 50) = 0 \tag{4.40}$$

for which the risk aversion coefficient increases to 0.0113. The second alternative would be to decrease the certainty equivalent (i.e., make it $25.75) yielding a risk aversion coefficient of 0.006942.

This approach has been applied by Dillon and Scandizzo (1978) to small farm owners and share croppers in northeastern Brazil. In this study, the researchers asked the decision makers to choose between "... two sets of simple yet reasonable mind experiments involving choice between risky and sure alternatives." In the first set of alternatives presented to the decision maker, total income was risky but the subsistence needs were assured. In the second set of alternatives, the subsistence requirements were also risky.

The researchers in this study used choices between a hypothetical, but realistic risky outcome versus a sure outcome. Given the initial response, the value of the preferred outcome was changed systematically until the subject stated that they were indifferent between the two outcomes. An example of the initial question is

> Which would you prefer — (A) to own a farm which gave you every year your family food requirements plus a net cash return of Cr$3500; or (B) to own a farm which in three years out of four gave you your family food requirements plus a net cash return of CR$4200 and in one year out of four gave you your family food requirements plus a net cash return of Cr$1400?
>
> (Dillon and Scandizzo 1978, p. 427).

Dillon and Scandizzo then regress the elicited risk aversion coefficients on a variety of socio-economic variables to analyze the effect that each factor has on risk aversion. They conclude that some but not all peasant farmers are risk averse. They also conclude that small owners and sharecroppers are more likely to be risk averse. Finally, they conclude that they "... believe our analysis has shown that it is possible via simple but purposive questioning to elicit meaningful information on peasant attitudes [toward risk] . . . ."

4.2.2. Equally Likely Risky Prospect and Finding Its Certainty Equivalent (ELCE)

Following the von Neumann and Morgenstern setup, Anderson, Dillon, and Hardaker (1977) propose another method to elicit risk by directly eliciting a utility function by finding the certainty equivalents of equally likely alternatives.

This process begins by assigning the expected utility at two endpoint outcomes. For example, consider a low income of $50,000 ($a = \$50,000$) and a high income of $100,000 ($b = \$100,000$). Next, assign a utility value at each endpoint ($U(50,000) = U(a) = 0$ and $U(100,000) = U(b) = 1$). Given this initial scenario, the researcher asks the decision maker how much they would be willing to give or take (basically what is the certainty equivalent) for a gamble that pays $50,000 with 0.5 probability and $100,000 with 0.5 probability (i.e., equally likely). Following the example in Table 4.3, assume that the decision maker responds that he is willing to take or give $70,000 for that gamble. Note that this response is consistent with

Table 4.3. Sequence of certainty equivalences for equally likely payoffs.

	Low	High	Certainty equivalent	Expected value
1	50,000	100,000	70,000	75,000
2	50,000	70,000	57,000	60,000
3	70,000	100,000	83,500	85,000
4	57,000	83,500	69,000	70,250

risk aversion since the certainty equivalent is less than the expected value, in this case \$75,000. For utility elicitation purposes, we assign the utility at the certainty equivalent a value of 0.50 ($U(70,000) = 0.50 \times U(50,000) + 0.50 \times U(100,000) = 0.50$ equal to the expected value of the endpoints).

To increase the precision of the estimated utility function, the analyst could then subdivide the range based on the first certainty equivalent. The second round of questions would begin by asking the decision maker how much he is willing to take or give for a gamble that pays \$50,000 with a probability of 0.5 and \$70,000 with a probability of 0.5 (i.e., taking the certainty equivalent from the first gamble and using it as an endpoint in the second gamble). Again following the result in Table 4.3, assume that the decision maker is willing to pay \$57,000 for this gamble. Again, this result is consistent with risk aversion because the certainty equivalent is less than the expected monetary value of this gamble at \$60,000. The utility value for this second gamble is 0.25 ($U(57,000) = 0.50 \times U(50,000) + 0.50 \times U(70,000) = 0.25$).

The entire ELCE sequence is represented generically in Table 4.4. Given the two endpoint outcomes: the minimum outcome a and the maximum outcome b, the researcher scaled the initial utilities to be $U(a) = 0$ and $U(b) = 1$. The researcher then asks the decision maker what value he is willing to take or give for the equally likely payoff of a or b. The decision maker's response is recorded as c. In the notation in Table 4.4, the payoff of c with certainty is denoted $(c; 1.0)$ while the payoff of a with 0.5 probability and b with 0.5 probability is denoted $(a, b; 0.5, 0.5)$. The statement that the decision maker is indifferent

Table 4.4.   Sequence of eliciting the certainty equivalents.

Step	Elicited certainty equivalent	Utility calculations
1	Initial scale	$U(a) = 0;\ U(b) = 1$
2	$(c; 1.0) \sim (a, b; 0.50)$	$U(c) = 0.50\,U(a) + 0.50\,U(b) = 0.50$
3	$(d; 1.0) \sim (a, c; 0.50)$	$U(d) = 0.50\,U(a) + 0.50\,U(c) = 0.25$
4	$(e; 1.0) \sim (c, b; 0.50)$	$U(e) = 0.50\,U(c) + 0.50\,U(b) = 0.75$
5	$(c'; 1.0) \sim (d, e; 0.50)$	$U(c') = 0.50\,U(d) + 0.50\,U(e) = 0.50$

Table 4.5.   Observed cardinal utility levels.

Money	Utility
50,000	0.00
57,000	0.25
70,000	0.50
69,000	0.50
83,500	0.75
100,000	1.00

between the payoff of c with certainty and the equally likely gamble that pays a or b is denoted $(c; 1.0) \sim (a, b; 0.50, 0.50)$. Note that $U(c) = 0.50 \times U(a) + 0.50 \times U(b) = 0.50$.

Table 4.5 presents the utility levels and monetary incomes associated with the responses in Table 4.3. These results can be used to estimate the utility function in a way that allows the researcher to estimate a risk aversion coefficient. Specifically, using the results in Table 4.5 as data, utility can be estimated as a function of income

$$U(Y) = -1.7789 + 4.4492 \times 10^{-05} Y - 1.6743 \times 10^{-10} Y^2.$$

$$(4.41)$$

The Arrow–Pratt absolute risk aversion coefficient for this result is then

$$R_A(Y) = -\frac{2(-1.6743 \times 10^{-10})}{4.4492 \times 10^{-05} - 2(1.6743 \times 10^{-10})Y}.$$ $$(4.42)$$

Intuitively, as more intervals are gathered, the precision of the estimated risk aversion coefficient improves. However, this conjecture assumes that each response is consistent with the decision maker's true risk aversion.

4.3.   Summary

The producer's attitudes toward risk are embodied in each individual's risk aversion. A standard representation for these risk preferences is the Arrow–Pratt risk aversion coefficient. However, this formulation is sensitive to the scale of the gamble facing the decision maker. This chapter presents the formal derivation of both the Arrow–Pratt absolute and relative risk aversion coefficients and discusses the transformations of scale for each. In addition, the chapter describes how risk aversion could be elicited from producers.

Chapter 5

Portfolio Theory and Decision Making Under Risk

The benefit to holding a portfolio of assets follows the old adage: "Don't put all of your eggs in one basket." The assumption is that by holding a collection of assets instead of holding the entire investment portfolio in a single asset the normal market fluctuations in individual investment returns will average out. However, does the mathematics support the old adage? This section examines the mathematical basis of portfolio theory. As a beginning point, define the "optimal portfolio" as that portfolio that minimizes the variance for a given expected return. Mathematically, letting z_i denote the share of asset i held in the portfolio, r_i denote the return on asset i, and $V(.)$ denote the variance, the problem of selecting the optimum portfolio can be written as

$$\min_z V \left(\sum_{i=1}^{12} z_i r_i \right)$$

$$s.t. \ \sum_{i=1}^{12} z_i \bar{r}_i \geq 0.03 \qquad (5.1)$$

$$\sum_{i=1}^{12} z_i = 1.00$$

where

$$V\left(\sum_{i=1}^{12} z_i r_i\right) = \sum_{t=1}^{12}\left(\sum_{i=1}^{4} z_i r_{it} - \sum_{i=1}^{4} z_i \bar{r}_i\right)^2$$

$$\bar{r}_i = \frac{1}{12}\sum_{t=1}^{12} r_{it}$$

(5.2)

and r_{it} denotes the observed return on investment i in year t. From the first equation in Eq. (5.1), the objective function is to minimize the variance of the rate of return on investment. The minimization criteria in the objective function (from the second equation in Eq. (5.1)) is consistent with the expected value–variance formulation discussed in Chapter 3. Specifically, by Proposition 2 of the Meyers (1987) formulation on page 73 of Chapter 3, utility is a decreasing function of the standard deviation (given that the distribution functions follow the location-scale formulation). Hence, portfolios with smaller variances are preferred to portfolios with larger variances. In addition, the constraint on expected income in Eq. (5.1) is consistent with Proposition 1 on page 73 of Chapter 3, which states that expected utility is an increasing function of the expected return. Alternatively, the optimal portfolio formulation in Eq. (5.1) can be justified by assuming that returns are normally distributed and preferences are negative exponential as depicted in Eq. (3.39). These formulations are typically referred to as expected value-variance formulations.

From an expected value–variance sense, one asset dominates another asset if it produces the same expected value for a lower variance. Consider the return data for Archer–Daniels–Midland, 3M, Hewlett-Packard, and John Deere presented in Table 5.1 taken from the WRDS (Wharton Business School, University of Pennsylvania) database. Using this data, the returns on 3M stock dominates both Archer–Daniels–Midland and Hewlett-Packard. However, the returns on 3M stock cannot be compared with John Deere stock returns because John Deere implies a higher risk than 3M. However, John Deere also has a higher rate of return.

Table 5.1. Portfolio example.

Date	Archer–Daniels–Midland	3M	Hewlett-Packard	John Deere	Portfolio
01/31/03	−0.0282	0.0101	0.0029	−0.0796	−0.0190
02/28/03	−0.0905	0.0119	−0.0896	−0.0194	−0.0192
03/31/03	−0.0092	0.0372	−0.0139	−0.0459	0.0030
04/30/03	0.0259	−0.0307	0.0482	0.1215	0.0277
05/30/03	0.0857	0.0086	0.1963	−0.0082	0.0368
06/30/03	0.0752	0.0198	0.0964	0.0515	0.0439
07/31/03	0.0210	0.0870	−0.0061	0.1112	0.0754
08/29/03	0.0601	0.0209	−0.0586	0.1128	0.0381
09/30/03	−0.0548	−0.0304	−0.0246	−0.0527	−0.0376
10/31/03	0.0946	0.1419	0.1524	0.1371	0.1386
11/28/03	0.0000	0.0063	−0.0255	0.0101	0.0023
12/31/03	0.0651	0.0758	0.0603	0.0660	0.0700
Average	0.0204	0.0299	0.0282	0.0337	0.0300
Variance	0.0035	0.0025	0.0073	0.0058	0.0011
Portfolio	0.0741	0.4997	0.1449	0.2813	1.0000

To demonstrate the risk reduction afforded by portfolios, solve the optimization problem specified in Eqs. (5.1) and (5.2). This solution yields the portfolio weights presented in the last row of Table 5.1. The results indicate that the largest share of the investment portfolio is held in 3M followed by John Deere and Hewlett-Packard. Note that by construction, the average return on the portfolio is 0.03%, which is slightly over the return of holding 3M alone, but the risk of the portfolio parametrized as the portfolio variance is 0.0011 is less than half the variance of the 3M stock. In fact, the variance of the portfolio is much smaller than the variance of any of the individual stocks. Hence, this empirical example appears to confirm the old adage about diversification. The remainder of this chapter develops the advantages to diversification more rigorously.

Note that the variance depicted in Eq. (5.2) is somewhat different than the standard method of moments or maximum likelihood formulation of variances. Starting from a matrix of N observed returns on k

returns ($X \in M_{N \times k}$), the variance matrix Σ is estimated as

$$\hat{\Sigma} = \frac{1}{N} X'X - \bar{x}'\bar{x}, \qquad (5.3)$$

where $\bar{x}$ is a $1 \times k$ vector of average returns. To justify this expression, the estimate of the variance matrix can be written as

$$\hat{\Sigma} = \frac{1}{N}(X - J_{1N}\bar{x})'(X - J_{1N}\bar{x})$$

$$= \begin{pmatrix} \frac{1}{N}\sum_{i=1}^{N}(x_{i1} - \bar{x}_1)(x_{i1} - \bar{x}_1) & \frac{1}{N}\sum_{i=1}^{N}(x_{i1} - \bar{x}_1)(x_{i2} - \bar{x}_2) \\[2mm] \frac{1}{N}\sum_{i=1}^{N}(x_{i2} - \bar{x}_2)(x_{i1} - \bar{x}_1) & \frac{1}{N}\sum_{i=1}^{N}(x_{i2} - \bar{x}_2)(x_{i2} - \bar{x}_2) \\[2mm] \vdots & \vdots \\[2mm] \frac{1}{N}\sum_{i=1}^{N}(x_{ik} - \bar{x}_k)(x_{i1} - \bar{x}_1) & \frac{1}{N}\sum_{i=1}^{N}(x_{ik} - \bar{x}_k)(x_{i2} - \bar{x}_2) \end{pmatrix}$$

$$\begin{pmatrix} \cdots & \frac{1}{N}\sum_{i=1}^{N}(x_{i1} - \bar{x}_1)(x_{ik} - \bar{x}_k) \\[2mm] \cdots & \frac{1}{N}\sum_{i=1}^{N}(x_{i2} - \bar{x}_2)(x_{ik} - \bar{x}_k) \\[2mm] \ddots & \vdots \\[2mm] \cdots & \frac{1}{N}\sum_{i=1}^{N}(x_{ik} - \bar{x}_k)(x_{ik} - \bar{x}_k) \end{pmatrix}, \qquad (5.4)$$

where J_{N1} is an $N \times 1$ vector of ones. To develop this linkage note

$$\hat{\Sigma} = \frac{1}{N}(X - J_{N1}\bar{x})'(X - J_{N1}\bar{x})$$

$$= \frac{1}{N}(X'X - X'J_{N1}\bar{x} - \bar{x}'J'_{N1}X + \bar{x}'J'_{N1}J_{N1}\bar{x}) \qquad (5.5)$$

where $X'J_{N1} = N\bar{x}'$ since

$$X'J_{N1} = \begin{pmatrix} \sum\limits_{i=1}^{N} x_{i1} \\ \vdots \\ \sum\limits_{i=1}^{N} x_{ik} \end{pmatrix} = \begin{pmatrix} N\bar{x}_1 \\ \vdots \\ N\bar{x}_k \end{pmatrix}. \qquad (5.6)$$

Simplifying Eq. (5.5) yields

$$\hat{\Sigma} = \frac{1}{N}(X'X - N\bar{x}'\bar{x} - N\bar{x}'\bar{x} + N\bar{x}'\bar{x}') = \frac{1}{N}X'X - \bar{x}'\bar{x} \qquad (5.7)$$

where the last term results from the fact that $J'_{N1}J_{N1} = N$.

Using this formulation, the variance of a weighted sum of normal random variables is defined as

$$\sigma_z^2 = z'\Sigma z = \sum_{j=1}^{k}\sum_{i=1}^{k} z_i z_j \sigma_{ij}, \qquad (5.8)$$

where z is a vector of portfolio weights (such as the last row in Table 5.1). Substituting the first expression from Eq. (5.5), Eq. (5.8) can be rewritten as

$$\sigma_z^2 = \frac{1}{N}z'(X - J_{N1}\bar{x})'(X - J_{N1}\bar{x})z$$

$$= \frac{1}{N}((X - J_{N1}\bar{x})z)'((X - J_{N1}\bar{x})z)$$

$$= \frac{1}{N}\sum_{j=1}^{k}\left(\sum_{i=1}^{N}\{z_j(x_{ij} - \bar{x}_j)\}^2\right)$$

$$= \frac{1}{N}\sum_{i=1}^{N}\left(\sum_{j=1}^{k}\{z_j(x_{ij} - \bar{x}_j)\}^2\right) \qquad (5.9)$$

which is the measure of variance given in Eq. (5.2).

5.1. The Expected Value — Variance Frontier

The expected value–variance frontier depicts the tradeoff between expected return on the portfolio and the implied risk of holding that portfolio (or the locus of points depicting the maximum expected return for any level of risk parametrized as the variance). This frontier can be derived several ways, but one popular formulation is to solve the portfolio problem specified as

$$\max_z \sum_{i=1}^{12} z_i \bar{r}_i$$

$$s.t. \ V\left(\sum_{i=1}^{12} z_i r_i\right) \leq \sigma^2 \tag{5.10}$$

$$\sum_{i=1}^{12} z_i = 1.00$$

using the data provided in Eq. (5.1). In the vernacular of mathematics, the formulation presented in Eq. (5.10) is the dual of the minimization problem posed in Eq. (5.1). Solving this mathematical programming problem for the variance of each individual asset [see the General Algebraic Modeling System (GAMS) code presented in Appendix E] yields the points in the expected value–variance frontier depicted in Fig. 5.1. Figure 5.2 compares the optimal portfolios from Eq. (5.10) and the expected value–variance results for the individual stocks from Table 5.1. The efficient portfolios (those portfolios that yield the greatest average returns from solution of Eq. (5.10)) are denoted with squares and joined with the dotted line. This dotted line forms the efficient expected value-variance frontier. The expected value–variance values of the original assets are denoted with diamonds. Note that three of the four expected value–variance combinations associated with the original assets lie inside (or to the lower right) of the efficient expected value–variance frontier. This implies that the efficient portfolio yields a higher expected return for the same level of risk (measured as the variance). Thus, the individual stocks are dominated by some portfolio

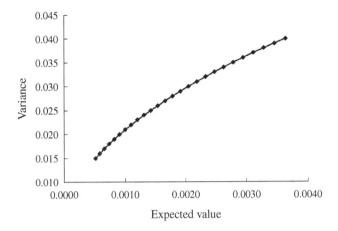

Fig. 5.1. Continuous expected value–variance frontier.

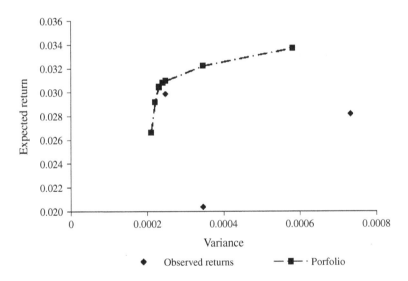

Fig. 5.2. Efficient portfolio for sample of stocks.

of stocks. Alternatively, holding a combination of stocks yields a higher expected return for the same level of variance as holding a given single stock. Hence, the results confirm our standard concepts of the gains to diversification.

Reformulate the problem using the normal distribution function:

$$\min_{z} \sigma_p^2(z) = z'\Sigma z$$
$$s.t. \; z'\bar{r} \geq \mu \qquad (5.11)$$
$$z \geq 0,$$

where the mean and variance of the returns are defined as in Eq. (5.3). In this example:

$$\bar{r} = \begin{bmatrix} 0.02041 \\ 0.02988 \\ 0.02818 \\ 0.03369 \end{bmatrix} \Sigma = \begin{bmatrix} 0.00316 & 0.00129 & 0.00355 & 0.00265 \\ 0.00129 & 0.00226 & 0.00129 & 0.00171 \\ 0.00355 & 0.00129 & 0.00670 & 0.00155 \\ 0.00265 & 0.00171 & 0.00155 & 0.00532 \end{bmatrix}$$

$$(5.12)$$

(see R code in Appendix F). Note that both the expected value–variance frontier depicted in Fig. 5.1 and the expected value–standard deviation frontier depicted in Fig. 5.2 are concave. Thus, relatively more risk must be incurred to increase the expected income from the portfolio.

5.2. A Simple Portfolio

The portfolio problem can also be recast within an expected utility framework. Letting $U(c_t)$ denote the utility of consuming c_i where t denotes a period in time. The problem then becomes how to allocate the individual's initial wealth (w_0) between alternative investments with a given price (p) to maximize the expected level of utility. The portfolio selection model can then be specified

$$\max_{c_1,h} E[U(c_0) + U(c_1)]$$
$$s.t. \; p'z = w_0 - c_1 \qquad (5.13)$$
$$c_{1j} = w_1 + x_j'z,$$

where x_j is the value of the vector of investments in state j. Thus, the decision maker chooses the portfolio of activities to maximize the expected utility over the set of possible states of nature at time period 1.

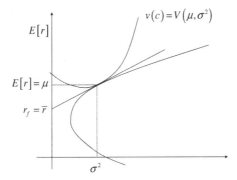

Fig. 5.3. Utility tangency and derivation of the efficient portfolio on the expected value–variance frontier.

This formulation will lead to the state-contingency formulation discussed later.

5.3. A Graphical Depiction of the Expected Value–Variance Frontier

Following on the above discussion, the expected value–variance frontier with a risk-free asset is depicted in Fig. 5.3. This general formulation will be used to construct the Capital Asset Pricing Model (CAPM) in Chapter 9. In essence, the portfolio decision is then restricted to a single efficient portfolio (in the CAPM model, this point is referred to as the market portfolio).

Theorem 5.1. *If an agent is strictly risk averse, then the optimal investment in the risky security is strictly positive, zero, or strictly negative if and only if the risk premium on the risky security is strictly positive, zero, or strictly negative.*

5.4. Mean–Variance versus Direct Utility Maximization

Due to various financial economic models such as the CAPM in the discussion of market models in Chapter 9, the finance literature relies on the use of expected value–variance decision rules rather than direct utility maximization. In addition, there is a practical aspect for stockbrokers who may want to give clients alternatives between efficient portfolios rather than attempting to directly elicit each individual's utility function (i.e., rather than attempting to elicit the investor's risk aversion coefficient using one of the approaches in Chapter 4, the stockbroker simply presents a list of expected value–variance alternatives and lets the decision maker choose between them). Consistent with this approach, the study by Kroll, Levy, and Markowitz (1984) examines the acceptability of the expected value–variance procedure. Specifically, they attempt to determine whether the expected utility maximizing choice is contained in the expected value–variance efficient set.

In order to compare the implications of maximizing expected utility against the expected value–variance formulation, begin by solving for the portfolio which maximizes expected utility formulated as

$$\max_{x} E[U[z]]$$

$$s.t. \sum_{i=1}^{n} z_i = 1 \tag{5.14}$$

$$z_i \geq 0.$$

Next, consider a reformulation of the expected value variance formulation presented from Eq. (5.10) consistent with the expected value–variance formulation based on normal returns and negative exponential preferences. Under these assumptions, the expected value–variance problem becomes

$$\max c'z - \frac{\rho}{2} z' \Omega z$$

$$s.t. \ z'1 = 1 \tag{5.15}$$

$$z_i \geq 0,$$

where ρ is the Arrow–Pratt absolute risk aversion coefficient.

Table 5.2 presents the optimum portfolios derived from direct maximization of expected utility as depicted in Eq. (5.14). Table 5.3

Table 5.2.   Optimal investment strategies with a direct utility maximization.

Utility	California	Carpenter	Chrysler	Conelco	Texas gulf	Avg. return	Std. Dev.
$-e^{-x}$	44.3	34.7	0.2	5.5	15.3	22.4	27.3
$x^{0.1}$	33.2	36.0		13.6	17.2	23.3	32.3
$x^{0.5}$		42.2		34.4	23.4	25.9	49.4
$\ln(x)$	37.9	34.8		11.1	16.2	23.1	29.4

Table 5.3.   Optimal E-V portfolios for various utility functions.

Utility	California	Carpenter	Chrysler	Conelco	Texas gulf	Avg. return	Std. Dev.
$-e^{-x}$	39.4	38.6		5.0	17.0	22.5	27.0
$x^{0.1}$	28.5	43.4		8.6	8.6	23.1	30.0
$x^{0.5}$		41.8		32.1	26.1	25.7	47.3
$\ln(x)$	32.9	41.8		7.4	18.7	22.9	28.9

presents the optimum portfolios selected by maximizing the expected value–variance formulation in Eq. (5.15) (using the utility functions presented in Table 5.3 to derive the Arrow–Pratt absolute risk aversion coefficient). Both the optimal portfolios and expected value–variance for each utility function are similar. To examine the difference in between the two portfolios, the root mean deviation[1] is taken as the measure of difference, the shares of utility maximizing portfolio in Table 5.2 as the true portfolio, and the expected value–variance is taken

---

[1]Here we define the root mean deviation (*rmd*) as

$$rmd = \left(\sum_{i=1}^{5} (\tilde{w}_i - w_i)^2 w_i \right)^{\frac{1}{2}},$$

where $\tilde{w}_i$ is the share of the portfolio invested in asset i under the expected value–variance formulation and w_i is the share of the portfolio invested in asset i under the expected utility formulation.

as an approximation. The relative error between the approximation and the true portfolio are computed to be $0.04, 0.07, 0.02$, and 0.05. Thus, the largest deviation is for the utility function $U(x) = x^{0.1}$, which is a form of the power utility function. Examining the differences between the average return on the portfolio and the standard deviation, the root mean squared error for average returns are $0.02, 0.04, 0.04$, and 0.04 while the root mean squared error for the portfolio standard deviations are $0.06, 0.40, 0.29$, and 0.09. Thus, the greatest differences appear to be in the standard deviations of portfolios.

5.5. Derivation of the Expected Value–Variance Frontier

To this point, this chapter has presented numerical results for the portfolio formulation using computer algorithms (such as GAMS). This section presents the analytical derivation of the optimal portfolios. These analytical results will be used throughout the remainder of this book. One caveat is that the analytical solutions must be kept tractable, and the derivations in the section do not impose nonnegativity restrictions on asset holdings. Hence, negative results are possible. These negative holdings would represent short-sales, which are permitted in the stock market under certain conditions, but infeasible if the expected value–variance formulation is used to depict many farm planning problems (i.e., the optimum selection of crops).

5.5.1. *Derivation without a Risk-Free Asset*

As a starting point, consider the general formulation without a risk-free asset. Beginning with the definition of the expected value and the variance

$$\bar{r} = z'\bar{r} + z_0 R = \sum_{i=0}^{N} z_i r_i$$

$$\sigma^2 = z'\Sigma z = \sum_{i=1}^{N} \sum_{j=1}^{N} z_i z_j \sigma_{ij}. \tag{5.16}$$

Using these definitions, the expected value–variance frontier without a risk-free (i.e., $z_0 = 0$) asset can be specified as

$$\min \frac{1}{2} z' \Sigma z$$
$$s.t. \ 1'z = 1 \tag{5.17}$$
$$\bar{r}'z = \mu,$$

where 1 is a conformable vector of ones. Forming the Lagrangian of Eq. (5.17)

$$L \equiv \frac{z' \Sigma z}{2} + \lambda(1 - 1'z) + \gamma(\mu - \bar{r}'z). \tag{5.18}$$

The first-order necessary conditions for Eq. (5.18) can be written in vector form as

$$\nabla_z L = \nabla_z \left(\frac{z' \Sigma z}{2} \right) - \lambda \nabla_z (1'z) - \gamma(\bar{r}'z), \tag{5.19}$$

where $\nabla_z L$ denotes a column vector of derivatives with respect to each z_i. First, focus on the variance of a portfolio with three possible investment alternatives. The variance for this scenario can be written as

$$z' \Sigma z = \begin{bmatrix} z_1 & z_2 & z_3 \end{bmatrix} \begin{bmatrix} \sigma_{11} & \sigma_{12} & \sigma_{13} \\ \sigma_{12} & \sigma_{22} & \sigma_{23} \\ \sigma_{13} & \sigma_{23} & \sigma_{33} \end{bmatrix} \begin{bmatrix} z_1 \\ z_2 \\ z_3 \end{bmatrix}$$

$$= \begin{bmatrix} z_1 & z_2 & z_3 \end{bmatrix} \begin{bmatrix} z_1 \sigma_{11} + z_2 \sigma_{12} + z_3 \sigma_{13} \\ z_1 \sigma_{12} + z_2 \sigma_{22} + z_3 \sigma_{23} \\ z_1 \sigma_{13} + z_2 \sigma_{23} + z_3 \sigma_{33} \end{bmatrix}$$

$$= z_1(z_1 \sigma_{11} + z_2 \sigma_{12} + z_3 \sigma_{13}) + z_2(z_1 \sigma_{12} + z_2 \sigma_{22} + z_3 \sigma_{23})$$
$$+ z_3(z_1 \sigma_{13} + z_2 \sigma_{23} + z_3 \sigma_{33}). \tag{5.20}$$

Hence, the gradient (or vector of derivatives) can be written as

$$\nabla_z z' \Sigma z = \begin{bmatrix} \dfrac{\partial z' \Sigma z}{\partial z_1} \\ \dfrac{\partial z' \Sigma z}{\partial z_2} \\ \dfrac{\partial z' \Sigma z}{\partial z_3} \end{bmatrix}$$

$$= \begin{bmatrix} (z_1\sigma_{11} + z_2\sigma_{12} + z_3\sigma_{13}) + z_1\sigma_{11} + z_2\sigma_{12} + z_3\sigma_{13} \\ (z_1\sigma_{12} + z_2\sigma_{22} + z_3\sigma_{23}) + z_1\sigma_{12} + z_2\sigma_{22} + z_3\sigma_{23} \\ (z_1\sigma_{13} + z_2\sigma_{23} + z_3\sigma_{33}) + z_1\sigma_{13} + z_2\sigma_{23} + z_3\sigma_{33} \end{bmatrix}$$

$$\nabla_z z' \Sigma z = \begin{bmatrix} 2(z_1\sigma_{11} + z_2\sigma_{12} + z_3\sigma_{13}) \\ 2(z_1\sigma_{12} + z_2\sigma_{22} + z_3\sigma_{23}) \\ 2(z_1\sigma_{13} + z_2\sigma_{23} + z_3\sigma_{33}) \end{bmatrix}. \tag{5.21}$$

Replicating these results using a little matrix calculus

$$\nabla_z z' \Sigma z = \Sigma z + \left(z' \Sigma \right)'. \tag{5.22}$$

Working the first term out

$$\Sigma z = \begin{bmatrix} \sigma_{11} & \sigma_{12} & \sigma_{13} \\ \sigma_{12} & \sigma_{22} & \sigma_{23} \\ \sigma_{13} & \sigma_{23} & \sigma_{33} \end{bmatrix} \begin{bmatrix} z_1 \\ z_2 \\ z_3 \end{bmatrix} = \begin{bmatrix} \sigma_{11}x_1 + \sigma_{12}x_2 + \sigma_{13}x_3 \\ \sigma_{12}x_1 + \sigma_{22}x_2 + \sigma_{23}x_3 \\ \sigma_{13}x_1 + \sigma_{23}x_2 + \sigma_{33}x_3 \end{bmatrix}$$

$$\tag{5.23}$$

while the second term becomes

$$z' \Sigma = \begin{bmatrix} z_1 & z_2 & z_3 \end{bmatrix} \begin{bmatrix} \sigma_{11} & \sigma_{12} & \sigma_{13} \\ \sigma_{12} & \sigma_{22} & \sigma_{23} \\ \sigma_{13} & \sigma_{23} & \sigma_{33} \end{bmatrix}$$

$$= [z_1\sigma_{11} + z_2\sigma_{12} + z_3\sigma_{13} \quad z_1\sigma_{12} + z_2\sigma_{22} + z_3\sigma_{23}$$

$$z_1\sigma_{13} + z_2\sigma_{23} + z_3\sigma_{33}]$$

$$\left(z' \Sigma \right)' = \begin{bmatrix} z_1\sigma_{11} + z_2\sigma_{12} + z_3\sigma_{13} \\ z_1\sigma_{12} + z_2\sigma_{22} + z_3\sigma_{23} \\ z_1\sigma_{13} + z_2\sigma_{23} + z_3\sigma_{33} \end{bmatrix}. \tag{5.24}$$

Putting the two halves together

$$\nabla_z z' \Sigma z = \Sigma z + (z' \Sigma)' = \begin{bmatrix} z_1 \sigma_{11} + z_2 \sigma_{12} + z_3 \sigma_{13} \\ z_1 \sigma_{12} + z_2 \sigma_{22} + z_3 \sigma_{23} \\ z_1 \sigma_{13} + z_2 \sigma_{23} + z_3 \sigma_{33} \end{bmatrix}$$

$$+ \begin{bmatrix} z_1 \sigma_{11} + z_2 \sigma_{12} + z_3 \sigma_{13} \\ z_1 \sigma_{12} + z_2 \sigma_{22} + z_3 \sigma_{23} \\ z_1 \sigma_{13} + z_2 \sigma_{23} + z_3 \sigma_{33} \end{bmatrix}$$

$$= \begin{bmatrix} 2(z_1 \sigma_{11} + z_2 \sigma_{12} + z_3 \sigma_{13}) \\ 2(z_1 \sigma_{12} + z_2 \sigma_{22} + z_3 \sigma_{23}) \\ 2(z_1 \sigma_{13} + z_2 \sigma_{23} + z_3 \sigma_{33}) \end{bmatrix}. \quad (5.25)$$

Turning to the constraints, the first constraint in Eq. (5.17) can be written as

$$1'z = \begin{bmatrix} 1 & 1 & 1 \end{bmatrix} \begin{bmatrix} z_1 \\ z_2 \\ z_3 \end{bmatrix} = z_1 + z_2 + z_3 \quad (5.26)$$

or in general

$$\alpha'z = \begin{bmatrix} \alpha_1 & \alpha_2 & \alpha_3 \end{bmatrix} \begin{bmatrix} z_1 \\ z_2 \\ z_3 \end{bmatrix} = \alpha_1 z_1 + \alpha_2 z_2 + \alpha_3 z_3. \quad (5.27)$$

Taking the gradient of the general form

$$\nabla_z \alpha'z = \begin{bmatrix} \alpha_1 \\ \alpha_2 \\ \alpha_3 \end{bmatrix}$$

$$\Rightarrow \nabla_z 1'z = \begin{bmatrix} 1 \\ 1 \\ 1 \end{bmatrix} \quad (5.28)$$

Therefore, returning to the Lagrangian

$$\nabla_z L = \Sigma z - \lambda 1 - \gamma \bar{r} = 0$$
$$\Sigma z = \lambda 1 + \gamma \bar{r} \tag{5.29}$$
$$z^* = \lambda \Sigma^{-1} 1 + \gamma \Sigma^{-1} \bar{r}$$

(remembering the $\frac{1}{2}$ term in Eq. (5.18)). Letting $\Sigma^{-1} = [\sigma^{ij}]$

$$z^* = \lambda \begin{bmatrix} \sigma^{11} & \sigma^{12} & \sigma^{13} \\ \sigma^{12} & \sigma^{22} & \sigma^{23} \\ \sigma^{13} & \sigma^{23} & \sigma^{33} \end{bmatrix} \begin{bmatrix} 1 \\ 1 \\ 1 \end{bmatrix} + \gamma \begin{bmatrix} \sigma^{11} & \sigma^{12} & \sigma^{13} \\ \sigma^{12} & \sigma^{22} & \sigma^{23} \\ \sigma^{13} & \sigma^{23} & \sigma^{33} \end{bmatrix} \begin{bmatrix} \bar{r}_1 \\ \bar{r}_2 \\ \bar{r}_3 \end{bmatrix}$$
$$= \begin{bmatrix} \lambda(\sigma^{11} + \sigma^{12} + \sigma^{13}) + \gamma(\sigma^{11}\bar{r}_1 + \sigma^{12}\bar{r}_2 + \sigma^{13}\bar{r}_3) \\ \lambda(\sigma^{12} + \sigma^{22} + \sigma^{23}) + \gamma(\sigma^{12}\bar{r}_1 + \sigma^{22}\bar{r}_2 + \sigma^{23}\bar{r}_3) \\ \lambda(\sigma^{13} + \sigma^{23} + \sigma^{33}) + \gamma(\sigma^{13}\bar{r}_1 + \sigma^{23}\bar{r}_2 + \sigma^{33}\bar{r}_3) \end{bmatrix}. \tag{5.30}$$

Using the constraint on the mean return to solve for γ^*

$$\bar{r}'z^* = \bar{r}'[\lambda \Sigma^{-1} 1 + \gamma \Sigma^{-1} \bar{r}] = \mu$$
$$\lambda \bar{r}' \Sigma^{-1} 1 + \gamma \bar{r}' \Sigma^{-1} \bar{r} = \mu. \tag{5.31}$$

To simplify the solution, define

$$B = \bar{r}' \Sigma^{-1} 1 = 1' \Sigma^{-1} \bar{r} \gtreqless 0$$
$$C = \bar{r}' \Sigma^{-1} \bar{r} > 0$$
$$\lambda B + \gamma C = \mu \tag{5.32}$$
$$\gamma^* = \frac{\mu - \lambda B}{C}.$$

Substituting this result back into the optimum portfolio in Eq. (5.29) yields

$$z^* = \left[\lambda \Sigma^{-1} 1 + \left(\frac{\mu - \lambda B}{C} \right) \Sigma^{-1} \bar{r} \right]. \tag{5.33}$$

Next, use the final first-order necessary condition to solve for λ

$$1'\left[\lambda \Sigma^{-1}1 + \left(\frac{\mu - \lambda B}{C}\right)\Sigma^{-1}\bar{r}\right] = 1$$

$$\lambda 1'\Sigma^{-1}1 + \left(\frac{\mu - \lambda B}{C}\right)1'\Sigma^{-1}\bar{r} = 1$$

$$1'\Sigma^{-1}1 \equiv A > 0$$

$$\lambda A + \left(\frac{\mu - \lambda B}{C}\right)B = 1 \tag{5.34}$$

$$\lambda AC + \mu B - \lambda B^2 = C$$

$$\lambda(AC - B^2) = C - \mu B$$

$$\lambda^* = \frac{C - \mu B}{AC - B^2}$$

substituting the result of Eq. (5.34) back into the solution γ for in Eq. (5.32)

$$\gamma^* = \frac{\mu - \left(\dfrac{C - \mu B}{AC - B^2}\right)B}{C}$$

$$= \frac{(AC - B^2)\mu - (C - \mu B)B}{C(AC - B^2)}$$

$$= \frac{\mu AC - B^2\mu - CB + \mu B^2}{C(AC - B^2)}$$

$$= \frac{\mu A - B}{AC - B^2}. \tag{5.35}$$

The stock portfolio using data Table 5.1 and fixing $\mu = 0.05$ is then

$$A = 522.2824$$
$$B = 13.1128$$
$$C = 0.4891$$
$$\lambda = \frac{C - \mu B}{AC - B^2} = -0.001995 \tag{5.36}$$
$$\gamma = \frac{\mu A - B}{AC - B^2} = 0.1557.$$

The optimal portfolio is then

$$z^* = \begin{bmatrix} -1.90738 \\ 0.98794 \\ 0.95442 \\ 0.96503 \end{bmatrix} \Rightarrow \sigma^2 = 0.005188. \qquad (5.37)$$

Given that A, B, and C are constants the researcher can solve for $\mu = 0.04$; $\lambda = -0.000412$, $\gamma = 0.09265$, and the optimal portfolio becomes

$$z^* = \begin{bmatrix} -0.94501 \\ 0.86070 \\ 0.53621 \\ 0.54749 \end{bmatrix} \Rightarrow \sigma^2 = 0.0033024. \qquad (5.38)$$

Table 5.4 depicts the σ^2, λ^*, γ^*, and z^* as μ varies between 0.01 and 0.15. The expected return–variance frontier for these solutions is then depicted in Fig. 5.4.

Table 5.4.   Optimal portfolio without a risk free asset.

μ	σ^2	λ^*	γ^*	ADM	3M	HP	JD
0.01	0.0033	0.00429	−0.09450	1.942	0.479	−0.716	−0.705
0.02	0.0021	0.00272	−0.03190	0.980	0.606	−0.298	−0.288
0.03	0.0021	0.00115	0.03060	0.017	0.733	0.119	0.130
0.04	0.0033	−0.00042	0.09320	−0.945	0.861	0.537	0.547
0.05	0.0058	−0.00200	0.15570	−1.907	0.988	0.954	0.965
0.06	0.0095	−0.00357	0.21830	−2.870	1.115	1.372	1.383
0.07	0.0145	−0.00514	0.28090	−3.832	1.242	1.790	1.800
0.08	0.0208	−0.00671	0.34340	−4.794	1.370	2.207	2.218
0.09	0.0283	−0.00828	0.40600	−5.757	1.497	2.625	2.635
0.10	0.0370	−0.00985	0.46850	−6.719	1.624	3.042	3.053
0.11	0.0470	−0.01142	0.53110	−7.682	1.751	3.460	3.470
0.12	0.0582	−0.01299	0.59370	−8.644	1.879	3.878	3.888
0.13	0.0707	−0.01456	0.65620	−9.606	2.006	4.295	4.305
0.14	0.0845	−0.01613	0.71880	−10.569	2.133	4.713	4.723
0.15	0.0995	−0.01770	0.78140	−11.531	2.260	5.130	5.140

Computed using R code presented in Appendix F.

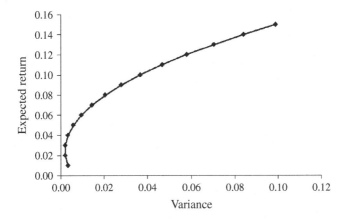

Fig. 5.4. Expected value-variance frontier without a risk-free asset.

5.5.2. *Derivation with a Risk-Free Asset*

If a risk-free asset is introduced into the portfolio, the efficient set of portfolios becomes a straight line between the risk-free asset and a tangency on the expected value-standard deviation frontier as depicted in Fig. 5.5 (note that the expected value-standard deviation frontier shares many of the same properties as the expected value–variance frontier).

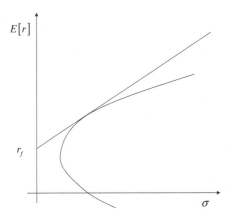

Fig. 5.5. Tangency between the expected value-standard deviation frontier with a risk-free asset.

Again, start by formulating the optimization problem as

$$\min \frac{1}{2} z' \Sigma z$$
$$s.t. \ (\bar{r} - R1)'z = \mu - R,$$

(5.39)

where R is the return on the risk-free asset. Forming the Lagrangian for the optimization problem in Eq. (5.39) yields

$$L = \frac{1}{2} z' \Sigma z - \gamma(\mu - R - (\bar{r} - R1)'z)$$
$$\nabla_z L = \Sigma z - \gamma(\bar{r} - R1) = 0$$
$$z^* = \gamma \Sigma^{-1}(\bar{r} - R1)$$

(5.40)

with $z_0^* = 1 - 1'z$ defined as the amount of wealth invested in the risk-free asset if $z_0^* > 0$ or borrowed if $z_0^* < 0$. Substituting this result into the constraint yields

$$(\bar{r} - R1)'[\gamma \Sigma^{-1}(\bar{r} - R1)] = \mu - R$$
$$\gamma[\bar{r}'\Sigma^{-1}\bar{r} - R\bar{r}'\Sigma^{-1}1 - R1'\Sigma^{-1}\bar{r} + R^2 1'\Sigma^{-1}1] = \mu - R.$$

(5.41)

From the derivation of the expected value–variance frontier without a risk-free asset, substitute $A = 1'\Sigma^{-1}1$, $B = 1'\Sigma^{-1}\bar{r}$, and $C = \bar{r}'\Sigma^{-1}\bar{r}$ into Eq. (5.41). Given these substitutions, the first-order condition becomes

$$\gamma[C - 2RB + R^2 A] = \mu - R$$
$$\gamma^* = \frac{\mu - R}{[C - 2RB + R^2 A]}.$$

(5.42)

The optimal portfolio is then defined as

$$z^* = \frac{\mu - R}{[C - 2RB + R^2 A]} \Sigma^{-1}(\bar{r} - R1).$$

(5.43)

The variance of the optimal portfolio becomes

$$
\begin{aligned}
\sigma^2 &= z'\Sigma z \\
&= \frac{(\mu - R)^2}{[C - 2RB + R^2 A]^2}[\Sigma^{-1}(\bar{r} - R1)]'\Sigma[\Sigma^{-1}(\bar{r} - R1)] \\
&= \frac{(\mu - R)^2}{[C - 2RB + R^2 A]^2}(\bar{r} - R1)'\Sigma^{-1}\Sigma\Sigma^{-1}(\bar{r} - R1) \\
&= \frac{(\mu - R)^2}{[C - 2RB + R^2 A]^2} \\
&\quad \times [\bar{r}'\Sigma^{-1}\bar{r} - R1'\Sigma^{-1}\bar{r} - R\bar{r}'\Sigma^{-1}1 + R^2 1'\Sigma^{-1}1] \\
&= \frac{(\mu - R)^2}{[C - 2RB + R^2 A]^2}[C - 2RB - R^2 A] \\
&= \frac{(\mu - R)^2}{[C - 2RB + R^2 A]}.
\end{aligned}
\tag{5.44}
$$

Looking ahead to the market valuation models for risk presented in Chapter 9

$$
\sigma = (\mu - R)\frac{1}{\sqrt{C - 2RB + R^2 A}}
$$

$$
\therefore \mu = R + (C - 2RB + R^2 A)^{\frac{1}{2}}\sigma \Rightarrow \text{Capital Market Line}
\tag{5.45}
$$

or the tradeoff between expected return and the standard deviation is linear.

Table 5.5 presents σ, γ^*, and z^* for μ between 0.02 and 0.15 and taking $R = 0.01$ for the mean vector and variance matrix from Eqs. (5.42) and (5.44). Since z_0^* is the net level of lending at the risk-free interest rate, the positive value of this variable at lower expected return implies that the individual is lending at the risk-free rate. However, at higher levels of expected returns, the decision maker is borrowing at the risk-free rate. Further, this behavior is independent of the short-sale of some of the assets in the optimal portfolio. Specifically, dividing each of the portfolios in Table 5.5 by $1 - z_0^*$ yields the

Table 5.5.   Optimal portfolio with a risk-free asset.

μ	σ	γ^*	ADM	3M	HP	JD	z_0^*
0.02	0.0004	0.03584	−0.413	0.263	0.215	0.218	0.717
0.03	0.0014	0.07168	−0.826	0.525	0.430	0.436	0.434
0.04	0.0032	0.10751	−1.240	0.788	0.645	0.654	0.152
0.05	0.0057	0.14335	−1.653	1.051	0.861	0.873	−0.131
0.06	0.0090	0.17919	−2.066	1.313	1.076	1.091	−0.414
0.07	0.0129	0.21503	−2.479	1.576	1.291	1.309	−0.697
0.08	0.0176	0.25087	−2.893	1.839	1.506	1.527	−0.979
0.09	0.0229	0.28670	−3.306	2.101	1.721	1.745	−1.262
0.10	0.0290	0.32254	−3.719	2.364	1.936	1.963	−1.545
0.11	0.0358	0.35838	−4.132	2.627	2.151	2.182	−1.828
0.12	0.0434	0.39422	−4.545	2.889	2.367	2.400	−2.110
0.13	0.0516	0.43006	−4.959	3.152	2.582	2.618	−2.393
0.14	0.0606	0.46589	−5.372	3.415	2.797	2.836	−2.676
0.15	0.0702	0.50173	−5.785	3.677	3.012	3.054	−2.959

Computed using R code presented in Appendix F.

percentage of the initial wealth held in each asset.

$$\frac{z^*}{1-z_0^*} = \begin{bmatrix} -1.4614 \\ 0.9290 \\ 0.7609 \\ 0.7715 \end{bmatrix}. \tag{5.46}$$

Figure 5.6 depicts the optimal mean–variance relationship including a risk-free asset compared with the optimal mean–variance relationship without a risk-free asset.

5.6.   Summary

The portfolio approach to decision making under risk has been a mainstay of risk analysis since Markowitz (1959). This analysis largely relies on the expected value–variance formulation. As demonstrated in Chapter 3, the expected value–variance objective functions are

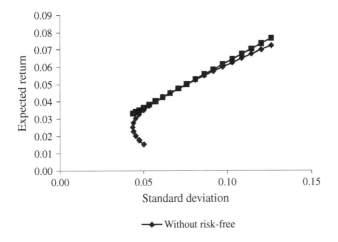

Fig. 5.6. Expected value–variance frontier with a risk-free asset.

theoretically consistent with the expected utility formulation proposed by von Neumann and Morgenstern. In addition, the results from this chapter indicate that the numerical results for the expected value–variance formulation of the portfolio approach are extremely close to the expected utility results. This chapter also presents the mathematical derivation of the expected value–variance frontiers with and without a risk-free asset.

Chapter 6

Whole Farm-Planning Models

Building on the preceding chapters, this chapter develops applied farm-planning models which incorporate risk. These models include direct application of the expected utility hypothesis drawing on the general formulation of the von Neumann–Morgenstern formulation, expected value–variance formulations at the whole farm level drawing heavily on the portfolio analysis presented in Chapter 5, and a variety of models that do not draw directly on the preceding theoretical development such as safety-first formulations, which maximize farm income given that a safety threshold is met.

6.1. Farm Portfolio Models

Starting from the discussion in Chapter 5, one formulation of the whole farm-planning model involves defining a farm plan (such as the selection of a portfolio of crops) using the expected value–variance approach. Specifically, assume that the decision maker chooses to minimize the variance associated with a crop rotation (or portfolio of crops) subject to a given level of income. The starting point for this formulation is a variance matrix for the rate of return per acre for the cropping alternatives under consideration

$$
\Omega = \begin{pmatrix}
924.41 & 458.52 & 202.22 & 135.22 \\
458.52 & 761.29 & 452.99 & 72.55 \\
202.22 & 452.99 & 490.11 & 109.09 \\
135.22 & 72.25 & 109.09 & 284.17
\end{pmatrix}
\tag{6.1}
$$

and a constraint matrix that depicts (1) the income constraint and (2) the portfolio constraint

$$Ax = b$$

$$\begin{pmatrix} 8.119 & 11.366 & 6.298 & 8.014 \\ 1.0 & 1.0 & 1.0 & 1.0 \end{pmatrix} \begin{pmatrix} x_1 \\ x_2 \\ x_3 \\ x_4 \end{pmatrix} = \begin{pmatrix} 7.0 \\ 1.0 \end{pmatrix} \quad (6.2)$$

$$8.119x_1 + 11.366x_2 + 6.298x_3 + 8.014x_4 = 7.000.$$

In Eq. (6.2), the expected return on the first crop alternative is 8.119 per acre, the expected return on the second crop alternative is 11.366 per acre, etc. In the second equation of Eq. (6.2), each acre of crop adds one acre to the overall acreage constraint. Therefore, the solution will yield crop shares in the optimal farm crop portfolio.

The portfolio choice problem can then be stated as

$$\min_{x} f(x) = x'\Omega x$$
$$s.t.\ Ax = b. \quad (6.3)$$

The optimal solution for this empirical model is

$$x = \begin{pmatrix} 0.11613 \\ -0.10666 \\ 0.39508 \\ 0.59546 \end{pmatrix}, \quad (6.4)$$

which yields a variance of 228.25 (see Appendix H for the GAMS code for this problem). Note that the optimum level for x_2 is negative. While a negative result is not a problem in designing a stock market portfolio (i.e., where a negative holding implies a short-sale), the negative value is problematic in the farm-planning model, where short-selling a crop is infeasible. Hence, most farm-planning model(s) impose nonnegativity restrictions on cropping levels.

The quadratic formulation of the farm-planning model originated with Freund (1956). To develop Freund's formulation start with the basic model of portfolio choice:

$$\min_{x} x'\Omega x$$
$$s.t. \; x'\mu > \Upsilon^*. \tag{6.5}$$

Freund showed that the expected utility of a normally distributed gamble given negative exponential preferences can be written as

$$E[U[x]] = \mu(x) - \frac{\rho}{2}\sigma^2(x), \tag{6.6}$$

(see pages 71–72 in Chapter 3). Extending this formulation to vector form, the problem can be expressed as

$$\max_{x} x'\mu - \frac{\rho}{x}x'\Omega x. \tag{6.7}$$

Given this formulation, Freund constructs a table of resource requirements for the farm level planning problem presented in Table 6.1 (note that Freund formulates the farm-planning problem on $100 of expected farm revenue rather than acres). The variance matrix is

Table 6.1. Freund's production constraints.

	Early Irish potatoes	Corn	Beef	Fall cabbage	Available
Land					
January–June	1.199	1.382	2.776	0.000	60.0
July–December	0.000	1.382	2.776	0.482	60.0
Production Capital					
Period 1	1.064	0.484	0.038	0.000	24.0
Period 2	−2.064	0.020	0.107	0.229	12.0
Period 3	−2.064	−1.504	−1.145	−1.229	0.0
Managerial Labor					
Period 1	5.276	4.836	0.000	0.000	799.0
Period 2	2.158	4.561	0.000	4.198	867.0
Period 3	0.000	4.561	0.000	13.606	783.0
Unit level of $100.0					
	1.000	1.000	1.000	1.000	

defined as

$$\Omega = \begin{pmatrix} 7304.69 & 903.89 & -688.73 & -182.05 \\ & 620.16 & -471.14 & 110.43 \\ & & 1124.64 & 750.69 \\ & & & 3689.53 \end{pmatrix}. \quad (6.8)$$

Letting x_1 be Irish potatoes, x_2 be corn, x_3 be beef, and x_4 be fall cabbage, the optimization problem is written as

$$\max_{x} \begin{pmatrix} 1 & 1 & 1 & 1 \end{pmatrix} \begin{pmatrix} x_1 \\ x_2 \\ x_3 \\ x_4 \end{pmatrix} - \frac{\rho}{2} \begin{pmatrix} x_1 \\ x_2 \\ x_3 \\ x_4 \end{pmatrix}' \Omega \begin{pmatrix} x_1 \\ x_2 \\ x_3 \\ x_4 \end{pmatrix}$$

subject to

$$1.199x_1 + 1.382x_2 + 2.776x_3 + 0x_4 \leq 60.0$$

$$0x_1 + 1.382x_2 + 2.776x_3 + 0.482x_4 \leq 60.0$$

$$1.064x_1 + 0.484x_2 + 0.038x_3 + 0.0x_4 \leq 24.0 \quad (6.9)$$

$$-2.064x_1 + 0.020x_2 + 0.107x_3 + 0.229x_4 \leq 12.0$$

$$-2.046x_1 - 1.504x_2 - 1.145x_3 - 1.229x_4 \leq 0.0$$

$$5.276x_1 + 4.836x_2 + 0x_3 + 0x_4 \leq 799.0$$

$$2.158x_1 + 4.561x_2 + 0x_3 + 4.198x_4 \leq 867.0.$$

Using a $\theta = 1/1,250.0$ the optimal solution under risk becomes

$$x = \begin{pmatrix} 10.29 \\ 26.76 \\ 2.68 \\ 32.35 \end{pmatrix}, \quad (6.10)$$

which conforms to Freund's original solution. The objective function for this optimum solution is \$5,383.08. Putting zero for θ yields an objective function of \$9,131.11 with an allocation of

$$x = \begin{pmatrix} 22.14 \\ 0.00 \\ 11.62 \\ 57.55 \end{pmatrix}. \quad (6.11)$$

(note that this solution uses a nonnegativity constraint for each activity). Question: How does the solution in Eq. (6.11) compare to the risk averse solution? Which crop makes the greatest gain? Which crop has the largest loss? The solution in Eq. (6.11) includes more fall cabbage, which is a more profitable crop, but also more risky. A second point is that although the objective function under risk aversion is $5,383.08, the expected income is $7,207.24.

6.1.1. *Gains to Diversification Using Certainty Equivalence*

The traditional formulation of the expected value–variance rules begins with the negative exponential utility function:

$$U[W(x)] = -e^{-\rho W(x)}. \tag{6.12}$$

In the previous discussion, this expected utility was rewritten under normality as

$$E[U(W\{x\})] = -e^{-\rho(\mu(x)-\frac{\rho}{2}\sigma^2(x))}. \tag{6.13}$$

Hence, the farm planning formulation typically maximizes

$$z = \mu(x) - \frac{\rho}{2}\sigma^2(x), \tag{6.14}$$

(as depicted in Eq. (6.9)). The implications of this objective function are actually much broader (Featherstone and Moss, 1990). Solving the negative exponential utility function for wealth

$$U[W^*(x)] = -e^{-\rho W^*(x)}$$
$$\ln[-U[W^*(x)] = -\rho W^*(x)] \tag{6.15}$$
$$W^*(x) = -\frac{1}{\rho}\ln[-U[W^*(x)]],$$

where $W^*(x)$ is the certainty equivalent of wealth as defined in Chapter 3. Substituting into the original integration (in Eq. (6.13)) yields

$$W^*(x) = \mu(x) - \frac{\rho}{2}\sigma^2(x). \tag{6.16}$$

Thus, the objective function of the expected value–variance formulation is actually the certainty equivalence.

Other implications of the equivalence between the objective function and the certainty equivalence are the interpretation of the shadow values of the constraint as changes in certainty equivalence. For example, given the original specification of the objective function, the shadow values of the second land constraint is 34.73 and the shadow value of the first capital constraint is 93.98. These values are then the price of each input under uncertainty.

6.1.2. *Extension to a Multiperiod Portfolio*

Historically, ownership of agricultural assets has been dominated by farmer equity and debt capital. Agriculture has typically not been able to attract capital in the form of stock ownership or limited partnerships. The implication of this form of ownership is increased variability in the return on equity to farmers (see Collins, 1985). A direct manifestation of the unwillingness of nonfarm investors to invest in agriculture can be seen in the unexplained premium on farm assets in the Capital Asset Pricing Model (see Barry, 1980). Moss, Featherstone, and Baker (1987) examine whether autocorrelation in the returns on farm assets versus other assets may explain the discrepancy.

Autocorrelation in farm returns refers to the tendency of increased returns to persist over time. Mathematically, this persistence can be represented as

$$\varepsilon_t = P\varepsilon_{t-1} + \mu_t, \tag{6.17}$$

where ε_t is an $n \times 1$ vector of asset returns, P is an $n \times n$ matrix of autocorrelation coefficients linking this period's returns with the returns in the preceding period, and μ_t is a vector of normal uncorrelated innovations. Given a vector of returns, the problem is to design the expected value–variance problem for holding a given portfolio of assets over several periods. Mathematically, this produces two problems: First, given the autoregressive structure of the problem, what is the expected return? From the simplistic perspective, the expected return for the holding period is now:

$$\tilde{\varepsilon}_t = k\bar{\varepsilon}_t \qquad (6.18)$$

where k is the number of periods held. However, a bigger problem is: What is the average income? In Moss, Featherstone, and Bakers' study, the average income is defined as the steady state income:

$$(I - AL)\varepsilon_t = \alpha_0$$
$$\varepsilon_t = (I - AL)^{-1}\alpha_0. \qquad (6.19)$$

If the eigenvalues of $(I - A)$ are between 1 and -1, the autoregressive representation is stationary.

A similar problem involves the variance matrix. Using the autoregressive estimation above, the variance matrix for the investment can be written as

$$
\Omega = \begin{pmatrix}
\Omega & P\Omega \\
\Omega P' & P\Omega P' + \Omega \\
\Omega P'P' & (P\Omega P' + \Omega)P' \\
\vdots & \vdots \\
\Omega (P')^{T-1} & (P\Omega P' + \Omega)(P')^{T-2}
\end{pmatrix}
$$

$$
\begin{pmatrix}
PP\Omega & \cdots & P^{T-1}\Omega \\
P(P\Omega P' + \Omega) & \cdots & P^{T-2}(P\Omega P + \Omega) \\
PP\Omega P'P' + P\Omega P' + \Omega & \cdots & P^{T-3}(PP\Omega P'P' + P\Omega P' + \Omega) \\
\vdots & \ddots & \vdots \\
(PP\Omega P'P' + P\Omega P' + \Omega)(P')^{T-3} & \cdots & \sum_{i=0}^{T-1} P^i\Omega(P')^i
\end{pmatrix}.
$$

$$(6.20)$$

Table 6.2. Mean and standard deviation of multiperiod formulation.

	Mean	Std. Dev.
Corporate bonds	1.24	6.71
Common stocks	6.01	19.29
Farm assets	4.65	3.87
Government bonds	0.55	6.71
Money	−3.06	7.14
Prime interest	1.81	20.74
Small stocks	8.64	29.83
Treasury bills	0.22	2.00

Empirically, this formulation implies the continuous time mean and standard deviation presented in Table 6.2. The correlation between the returns for the single year, three year, and five year holding periods is presented in Table 6.3 (i.e., the Ω has been normalized so that the diagonal element is one).

6.2. Minimize Total Absolute Deviation

Prior to the widespread availability of practical computer code to solve nonlinear programming formulations, the quadratic nature of the expected value–variance frontier limited the usefulness of Freund's formulation. This limitation spawned several specifications of the farm problem which approximated the expected value–variance solution. One such approximation was Hazell's (1971) minimize total absolute deviation (MOTAD).

Hazell's approach is twofold. He first reviews expected value–variance; suggesting that it is a good methodology under certain assumptions. Then, he raises two difficulties. The first difficulty is the availability of code to solve the quadratic programming problem implied by expected value–variance. The second problem is the estimation problem. Specifically, the data required for expected value–variance are the mean and the variance matrix. However, the variance matrix is an artifact of the assumption of normality.

Table 6.3.   Variance matrix for various holding periods.

	Corp. bonds	Common stocks	Farm assets	Gov. bonds	Money	Prime interest	Small stocks	Treasury bills
One Year Holding Period								
Corporate bonds	1.0000							
Common stocks	−0.4939	1.0000						
Farm assets	−0.9823	0.5543	1.0000					
Gov. bonds	0.9984	−0.5033	−0.9801	1.0000				
Money	0.8598	−0.4700	−0.8265	0.8650	1.0000			
Prime interest	0.8848	−0.5825	−0.9056	0.8865	0.8754	1.0000		
Small stocks	0.9273	−0.2194	−0.9041	0.9177	0.7268	0.7684	1.0000	
Treasury bills	0.5664	−0.6234	−0.6406	0.5686	0.5543	0.7978	0.7978	1.0000
Three Year Holding Period								
Corporate bonds	1.0000							
Common stocks	−0.4939	1.0000						
Farm assets	−0.9823	0.5543	1.0000					
Gov. bonds	0.9984	−0.5033	−0.9801	1.0000				
Money	0.8598	−0.4700	−0.8265	0.8650	1.0000			
Prime interest	0.8848	−0.5825	−0.9056	0.8865	0.8754	1.0000		
Small stocks	0.9273	−0.2194	−0.9041	0.9177	0.7268	0.7684	1.0000	
Treasury bills	0.5664	−0.6234	−0.6406	0.5686	0.5543	0.7978	0.7978	1.0000

(*Continued*)

Table 6.3. (*Continued*)

	Corp. bonds	Common stocks	Farm assets	Gov. bonds	Money	Prime interest	Small stocks	Treasury bills
Five Year Holding Period								
Corporate bonds	1.0000	0.0548	−0.9854	0.9987	0.8086	0.8559	0.9570	−0.4795
Common stocks		1.0000	0.0225	0.0326	−0.0909	−0.2253	0.2720	−0.5033
Farm assets			1.0000	−0.9852	−0.7793	−0.8961	−0.9286	0.3740
Gov. bonds				1.0000	0.8197	0.8649	0.9461	−0.4585
Money					1.0000	0.7827	0.6871	−0.3222
Prime interest						1.0000	0.7404	−0.0469
Small stocks							1.0000	−0.4586
Treasury bills								1.0000

The crux of the estimation problem is that the covariance terms in the expected value–variance formulation are estimated by

$$\sigma^2 = \sum_{j=1}^{n} \sum_{k=1}^{n} x_j x_k \left[\frac{1}{s-1} \sum_{h=1}^{s} (c_{hj} - g_j)(c_{hk} - g_k) \right],$$ (6.21)

where x_j is the level of activity j in the portfolio, c_{hj} is the observed return on asset j at time h, g_j is the expected return on asset j, s is the number of observations, and n is the number of assets. This expression can be reformulated as

$$\sigma^2 = \frac{1}{s-1} \sum_{h=1}^{s} \left[\sum_{j=1}^{n} c_{hj} x_j - \sum_{j=1}^{n} g_j x_j \right]^2.$$ (6.22)

Hazell suggests replacing this objective function with the mean absolute deviation

$$A = \frac{1}{s} \sum_{h=1}^{s} \left| \sum_{j=1}^{n} (c_{hj} - g_j) x_j \right|.$$ (6.23)

Thus, instead of minimizing the variance of the farm plan subject to an income constraint, the farm is assumed to minimize the absolute deviation subject to an income constraint. Another formulation for this objective function is to let each observation h be represented by a single row

$$y_h = \sum_{j=1}^{n} (c_{hj} - g_j) x_j$$

$$y_h^+ - y_h^- = \sum_{j=1}^{n} (c_{hj} - g_j) x_j,$$ (6.24)

where y_h is the deviation from average. This deviation can be divided into positive deviations from the average, y_h^+, and negative deviations from the average, y_h^-. The complete mathematical programming problem can then be written as

$$\min_x sA = \sum_{h=1}^{s} (y_h^+ + y_h^-)$$

$$s.t. \sum_{j=1}^{n} (c_{hj} - g_j)x_j - y_h^+ + y_h^- = 0$$

$$\sum_{j=1}^{n} g_j x_j = \lambda \qquad (6.25)$$

$$\sum_{j=1}^{n} a_{ij} x_j \leq b_i.$$

A last modification comes from refocusing on the negative deviations. Thus, the problem becomes:

$$\min_x sA = \sum_{h=1}^{s} y_h^-$$

$$s.t. \sum_{j=1}^{n} (c_{hj} - g_j)x_j + y_h^- \geq 0$$

$$\sum_{j=1}^{n} g_j x_j = \lambda \qquad (6.26)$$

$$\sum_{j=1}^{n} a_{ij} x_j \leq b_i.$$

A numerical example of MOTAD is presented in Table 6.4. Given this data, the mathematical formulation of the MOTAD problem from Eq. (6.26) can be expressed as:

$$\min_x y_1^- + y_2^- + y_3^- + y_4^- + y_5^- + y_6^-$$

$$x_1 + x_2 + x_3 + x_4 \leq 200$$

$$25x_1 + 36x_2 + 27x_3 + 87x_4 \leq 10000$$

$$-x_1 + x_2 - x_3 + x_4 \leq 0$$

Table 6.4.   Hazell's Florida farm.

	Carrots	Celery	Cucumbers	Peppers
1	292	−128	420	579
2	179	560	187	639
3	114	648	366	379
4	247	544	249	924
5	426	182	322	5
6	259	850	159	569
Average	253	443	284	516
1	39	−571	136	63
2	−74	117	−97	123
3	−139	205	82	−137
4	−6	101	−35	408
5	173	−261	38	−511
6	6	407	−125	53

$$39x_1 - 571x_2 + 136x_3 + 63x_4 + y_1^- \geq 0$$

$$-74x_1 + 117x_2 - 97x_3 + 123x_4 + y_2^- \geq 0$$

$$-139x_1 + 205x_2 + 82x_3 - 137x_4 + y_3^- \geq 0$$

$$-6x_1 + 101x_2 - 35x_3 + 408x_4 + y_4^- \geq 0$$

$$173x_1 - 261x_2 + 38x_3 - 511x_4 + y_5^- \geq 0$$

$$6x_1 + 407x_2 - 125x_3 + 53x_4 + y_6^- \geq 0$$

$$253x_1 + 443x_2 + 284x_3 + 516x_4 = \lambda. \qquad (6.27)$$

6.3.   Focus-Loss

Another formulation for decision making under risk is the Focus-Loss model, which was proposed by Shackle (1949, 1961) and it was originally applied to the farm-planning model by Boussard and Petit (1967). Boussard and Petit describe the logic behind the model succinctly

> We assume that farmers choose, among various possible actions, the one which will maximize the expected gains, provided that the

possibility of ruin is so small that it can be neglected. (Boussard and Petit, 1967, p. 871)

Two factors make Focus-Loss acceptable. First, like Hazell's MOTAD, the Focus-Loss problem is solvable using linear programming. Second, Focus-Loss has a direct appeal in that it focuses attention on survivability.

The first step in the Focus-Loss methodology is to define the maximum allowable loss

$$L = E(z) - z_c = \sum_{j=1}^{n} E(c_j)x_j - E(F) - z_c, \qquad (6.28)$$

where L is the maximum allowable loss, $E(z)$ is the expected income for the firm, z_c is the required cash income, $E(c_j)$ is the expected income from each crop j, x_j is the level of the jth crop (activity), and $E(F)$ is the expected level of fixed cost.

Given this definition, the next step is to define the maximum deficiencies or losses arising from activity j.

$$r_j = E(c_j) - r_j^*, \qquad (6.29)$$

where r_j^* is the worst expected outcome. For example, a crop failure may give an r_j of $-\$100$, which would represent your planting cost. Given this potential loss, the Focus-Loss scenario is based on restricting the largest expected loss to be above some stated level

$$r_r x_j \leq \frac{L}{k}. \qquad (6.30)$$

In the Anderson, Dillon and Hardaker example

$$\max_{x} 72x_1 + 53.4x_2 + 88.8x_3 - 200$$

$$x_1 + x_2 + x_3 \leq 12$$

$$x_1 + x_3 \leq 8$$

$$30x_1 + 20x_2 + 40x_3 \leq 400$$

$$5x_1 + 5x_2 + 8x_3 \leq 80$$

$$60x_1 - \frac{1}{3}L \leq 0$$

$$44.5x_2 - \frac{1}{3}L \leq 0$$

$$74x_3 - \frac{1}{3}L \leq 0. \tag{6.31}$$

The choice of $k = 3$ is somewhat arbitrary. Two points about the Focus-Loss: allowing $L \to -\infty$, the Focus-Loss solution is the profit maximizing solution, and L can become large enough to make the linear programming problem infeasible.

Boussard and Petit implicitly determine $1/k$ with the statement that "... to express the condition that the focus-loss on one crop cannot exceed the fraction $1/k$ of the total loss ..." (p. 874). The question is then whether k is simply defined as the number of crop alternatives? A better justification for k could involve setting k from the notion that

$$r_j^* = \mu_j - t_p\sigma_j. \tag{6.32}$$

Thus, if we let t_p be -1.96, the maximum loss would be $1.96\sigma_j$

$$(\sigma_j t_p)x_j \leq \frac{L}{k}. \tag{6.33}$$

Hence, Focus-Loss is another linear farm-planning model that allows for the incorporation of risk. It is an open question whether its lexicographic nature which departs significantly from expected utility is a strength or a weakness. As noted by Boussard and Petit, this specification eliminates the need to elicit risk aversion coefficients and eliminates the normality debates.

6.4. Target MOTAD

The target MOTAD model is a two-attribute risk and return model. Return is measured as the sum of the expected return of each activity multiplied by the activity level. Risk is measured as the expected sum of the negative deviations of the solution results from a target–return level. Risk is then varied parametrically so that a risk–return frontier can be traced out. The target MOTAD model is stated as

$$\max_{x} E(z) = \sum_{j=1}^{n} c_j x_j$$

$$s.t. \sum_{j=1}^{n} a_{ij} x_j \leq b_i$$

$$T - \sum_{j=1}^{n} c_{rj} x_j - y_r \leq 0 \tag{6.34}$$

$$\sum_{r=1}^{n} p_r y_r = \lambda,$$

where x_j is the activity level for crop j, c_j is the expected return on crop j, a_{ij} is the technical coefficient in column i of row j, b_i is the right hand side of that technical row, c_{rj} is the rth outcome for activity j, T is the target loss, y_r is the transfer of the negative deviation, and λ is the target loss. The decision process can then be expressed as a locus of points, where the whole farm plan maximizes expected income subject to a target level of negative deviation.

6.5.   Direct Utility Maximization

To this point, this chapter has discussed several alternatives to direct utility maximization which were based on efficiency criteria (as in the case of expected value–variance) or some *ad hoc* specification of risk aversion as in the case of focus-loss. This section presents the application of a discrete form of expected utility analysis.

To demonstrate direct utility maximization, consider the scenario where a Midwestern farmer with 1,280 acres wants to select the optimum portfolio of corn, soybeans, and wheat that maximizes the expected utility of terminal wealth. The return data for this formulation are presented in Table 6.5. To flush out the formulation, assume that the expected annual profit for the farmer is $271,782. Amortizing this amount into perpetuity using a discount rate of 15% yields a total value of $1,811,880. Assuming the debt-to-asset position of the farm is 60%, the value of the asset represents equity of $724,752 and debt

Table 6.5.   Data for direct utility maximization.

	Corn	Soybeans	Wheat
Observation 1	176.24	94.81	97.09
Observation 2	232.93	114.39	120.18
Observation 3	273.01	144.50	108.75
Observation 4	221.59	114.32	87.48
Observation 5	−7.87	97.22	100.46
Observation 6	247.59	126.41	108.34
Observation 7	226.79	113.49	98.16
Observation 8	250.11	123.27	107.60
Observation 9	255.99	136.15	102.81
Observation 10	246.91	131.04	104.68
Average	212.33	119.56	103.56

of $1,087,130. Assume an interest rate of 12.5% yields an annual cash flow requirement of $135,891 to cover the interest payments. Further assuming a family living requirement of $50,000 yields a minimum cash requirement of $185,891.

Next, consider the result for Observation 1 in Table 6.5. Subtracting the cash requirement from initial equity ($724,752 − $185,891 = $538,861) yields a wealth constraint of

$$W_i = 176.24x_1 + 94.81x_2 + 97.09x_3 + 538861, \qquad (6.35)$$

where x_1 is the number of acres planted in corn, x_2 is the number of acres planted in soybeans, and x_3 is the number of acres planted in wheat. A model of crop choice based on the expected value of the power utility function can then be defined as

$$\max_x \frac{1}{10}\frac{W_1^b}{b} + \frac{1}{10}\frac{W_2^b}{b} + \cdots + \frac{1}{10}\frac{W_{10}^b}{b}$$
$$x_1 + x_2 + x_3 \leq 1280$$
$$-176.2x_1 - 93.8x_2 - 97.1x_3 + W_1 = 538,861 \qquad (6.36)$$
$$-232.9x_1 - 114.4x_2 - 120.2x_3 + W_2 = 538,861$$
$$\vdots$$
$$-246.9x_1 - 131.0x_2 - 104.7x_3 + W_{10} = 538,861.$$

Table 6.6. Expected utility results.

Risk aversion	W	$W/1000$
$r = -1.0$	$-1.233\text{E-}06$	$-1.233\text{E-}03$
$r = -2.0$	$-7.600\text{E-}13$	$-7.609\text{E-}07$
$r = -3.0$	$-6.257\text{E-}19$	$-6.27\text{E-}10$

Table 6.7. Portfolio from expected utility.

r	x_1	x_2	x_3	μ	σ
-0.001	929.12	350.88	0.00	239,231.70	75,923.56
-0.1	882.59	397.41	0.00	234,915.10	72,865.46
-1.0	532.69	747.31	0.00	202,455.70	50,130.77
-10.0	4.93	0.00	284.59	30,517.50	2,479.89

In order to efficiently numerically solve this problem, the objective function (expected utility) must also be scaled. Specifically, assuming an average payoff, the terminal wealth is presented in Table 6.6. In the second column, the objective number is much smaller, which will lead to difficulties in the numerical optimization. As a result, the analyst substitutes the wealth defined in thousands of dollars in the third column which helps scale the optimization problem. In addition, the objective function is also directly scaled in an attempt to get an optimal value of the objective function to be around -1 to -10. The optimal solutions are then given by the crop rotations presented in Table 6.7.

6.6. Discrete Sequential Stochastic Programming

Target MOTAD, direct expected utility, and even MOTAD begin to develop the concept of constraints being stochastic or met with some level of probability. In target MOTAD, income under a certain state exceeds the target level of income with some probability. In direct expected utility maximization, the level of wealth transferred to the objective function was represented by a constraint which had some level of probability. MOTAD minimizes the expected negative deviations which implied stochastic constraints. However, in each of these cases, the primary impact of stochastic constraints was on the objective function or some threshold level of risk (as was the case in target MOTAD).

An alternative that focuses on the effect of risk on the constraints can be found in the Discrete Sequential Stochastic Programming (DSSP) formulation which originated in Cocks (1968). This formulation focuses on decision processes which are strung out over a discrete number of decision periods. The general formulation of the decision problem is depicted in Fig. 6.1. At a discrete point in the future, the farmer has to make a decision, for example a stocking rate on cattle. Given this first round decision and a random outcome, such as rainfall, there is then a subsequent decision to be made. For example, whether to sell cattle or buy feed. Each state occurs with a given level of probability and each "node" can contribute to the objective function.

A mathematical formulation

$$\max_{x} -c_1x_1 - c_2x_2 - c_3x_3 - c_4x_4 + c_5x_5 - c_6x_6 + c_7x_7 - c_8x_8 - c_9x_9 = 0$$
$$-s_{11}x_1 + u_{11}x_2 - f_{11}x_3 = 0$$
$$-s_{21}x_1 + u_{11}x_2 - f_{21}x_4 = 0$$
$$- x_2 + x_5 - x_6 + x_7 = 0$$
$$-s_{12}x_1 + u_{21}x_7 - f_{12}x_8 = 0$$
$$-s_{22}x_1 + u_{22}x_7 - f_{22}x_9 = 0.$$

$$(6.37)$$

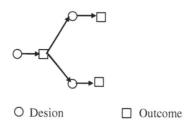

○ Desion   □ Outcome

Fig. 6.1.   Depiction of the DSSP state space.

In this model, x_1 represents the acres of wheat planted, x_2 is the number of stockers purchased, x_3 the tons purchased under outcome 1, and x_4 the tons of feed under outcome 2. The first two equations, then, simply balance the feed requirements under each state of nature. For example, if there is good rainfall in state 1, then more grazing will be produced by the wheat (x_2), and less feed will have to be purchased than in state 2. c_1 and c_2 are then the cost of feed in each state weighted by the probability of that state. The third equation then transfers the cattle purchased into the next decision period. x_5 is a variable modeling the number of stockers sold, while x_6 models any additional stockers purchased. The total number of stockers in the next production period is x_7. Given the number of cattle transferred into the next period, the feed balance relationships determine the level of feed that must be purchased.

6.7.   Chance-Constrained Programming

The DSSP problem above assumes that the possible outcomes can be represented in a finite number of states, although several pieces of applied research have examined the efficiency of approximating the moments of a continuous distribution with a finite number of points. An alternative would be to constrain the probability. For example, assume that you want to constrain the probability that profit will be less than a fixed level T (to borrow the target MOTAD concept). Mathematically, this constraint becomes:

$$P[X < x^*] \leq T. \tag{6.38}$$

Under normality, we can transform this constraint via the confidence interval:

$$x'\mu - \alpha_{0.05}\sqrt{x'\Omega x} \geq T$$
$$\alpha_{0.05} \geq \frac{x'\mu}{\sqrt{x'\Omega x}} - \frac{T}{\sqrt{x'\Omega x}}. \tag{6.39}$$

An alternative way to write this constraint is

$$\sum_j x_{ij}\mu_{ij} + D^* \left[\left(\sum_j \sigma_{ij}^2 x_j^2 \right)^{1/2} \right] \le b_i. \qquad (6.40)$$

Another way to envision this formulation is that while the typical assumption is that the constraints are known with certainty, in the chance-constrained formulation the coefficients of the constraints are random. The chance-constrained formulation was originally proposed by Charnes and Cooper (1959). Paris and Easter (1985) apply the general framework to Australian agriculture.

The typical mean–variance crop selection model is expressed as

$$\max_x E(p)'x - \frac{\rho}{2}x'\Sigma x$$
$$s.t.\ Ax \le b \qquad (6.41)$$
$$x \ge 0.$$

As discussed in the previous section, this specification adjusts for risk in the objective function, but only requires the constraints to be met at the mean.

An extension of this model involves appending a term on the constraints which accounts for risk in the constraints. Specifically, rephrasing the profit function as

$$\pi = p'x - s'v = d'z$$
$$d = \begin{pmatrix} p \\ s \end{pmatrix} \qquad (6.42)$$
$$z = \begin{pmatrix} x \\ y \end{pmatrix}.$$

Expected profit is then distributed

$$\pi \sim N(E(\pi), z'\Sigma z) \sim N(E(p)'x - E(s)y', x'\Sigma_p x + y'\Sigma_s y)$$
$$\Sigma = \begin{pmatrix} \Sigma_p & 0 \\ 0 & \Sigma_s \end{pmatrix}. \qquad (6.43)$$

In this case, the vector s represents the shadow values of the resource constraint. Thus, $s'v$ is the "cost" of using the resources.

This specification gives rise to a related pair of mathematical programming models. The primal

$$\max_{x} E(p)'x - \frac{\rho}{2}x'\Sigma_p x - \frac{\rho}{2}y'\Sigma_s y$$

$$Ax - \rho\Sigma_s \leq E(s) \tag{6.44}$$

$$y \geq 0, \quad x \geq 0$$

and the dual

$$\min_{x} E(s)'y + \frac{\rho}{2}y'\Sigma_s y + \frac{\rho}{2}x'\Sigma_p x$$

$$A'y + \rho\Sigma_p x \geq E(p) \tag{6.45}$$

$$y \geq 0, \quad x \geq 0.$$

This specification is consistent with chance-constrained programming. Specifically, maximizing the primal above can be viewed as maximizing the certainty equivalent of a risky revenue subject to the constraint that the probability of the marginal value of the constraints is less than a given critical level with some probability. Mathematically,

$$\Pr[y'Ax - y's \leq 0] \geq \alpha$$

$$\Pr\left[\frac{(-s'x + E(s)'y)}{(y'\Sigma_s y)^{\frac{1}{2}}} \leq \tau_s\right] = \alpha \tag{6.46}$$

$$\Pr[E(s)'y - \tau_s(y'\Sigma_s y)^{\frac{1}{2}} \leq s'y] = \alpha.$$

Portions of this derivation are dependent on the Kuhn–Tucker conditions. Specifically, given the augmented objective function:

$$L = E(p)'x - \frac{\rho}{2}x'\Sigma_p x - \frac{\rho}{2}y'\Sigma_s y + y'[E(s) + \rho\Sigma_s y - Ax]. \tag{6.47}$$

The Kuhn–Tucker conditions for a maximum become

$$\frac{\partial L}{\partial x} = E(p) - \rho\Sigma_p x - A'y \leq 0$$

$$x'\frac{\partial L}{\partial x} = x'E(p) - \rho x'\Sigma_p x - x'A'y = 0$$

$$\frac{\partial L}{\partial y} = E(s) + \rho \Sigma_s y - Ax \geq 0$$

$$y'\frac{\partial L}{\partial y} = y'E(s) + \rho y' \Sigma_s y - y'Ax = 0. \qquad (6.48)$$

6.8. Interpreting Shadow Values from Risk Programming Models

Several variants of mathematical programming that involve direct optimization of expected utility have been discussed. Now consider two specifics about the approach. The first involves interpretation of the dual variables from the optimization process and the second involves the parametrization of the probability space. In the expected value–variance procedures, certainty equivalence formulation of the expected value–variance problem from Featherstone and Moss (1990) was covered. This formulation allowed for the interpretation of the shadow values as changes in the certainty equivalent associated with a one-unit change in the right-hand side of the resource constraint. However, in the direct utility maximization problem, there is no such direct equivalence with a monetary value.

Preckel, Featherstone, and Baker (1987) propose a simple chain rule formulation used to derive a monetary value from the relationship between the shadow prices. Take a simple expected utility problem:

$$\max -\frac{1}{3}\exp\left(-\left[x_1^{0.5}x_2^{0.5} - d\right]\right) - \frac{2}{3}\exp\left(-\left[x_1^{0.5}x_2^{0.5} - d\right]\right)$$

$$s.t.\ cx_1 - d = 0 \qquad (6.49)$$

$$x_2 \leq 4$$

$$x_1, x_2 \geq 0.$$

In this problem, maximize the expected utility of profit for a Cobb–Douglas production function.

In this scenario, you have a stochastic output price, a fixed level of input 2, and a variable level of input 1. The price of input 1 is c. Each shadow price yields the number of additional utils provided by each unit increase in the right-hand side. This solution is not very satisfying.

However, noting that the first constraint is denominated in dollars, the inverse of this shadow value gives the dollars per util. Thus, the value of the second constraint can be derived by dividing the shadow value of the second constraint by the shadow value of the first constraint.

More rigorously,

$$\max f(x)$$
$$s.t.\, g_i(x) = b. \tag{6.50}$$

Writing the optimal level of x as a function of the right-hand side and substituting this into the Lagrangian we derive

$$L(x, \pi) = f(x(b)) - \sum_{i=1}^{m} \pi_i(b)[g_i(x(b)) - b_i]. \tag{6.51}$$

Totally differentiating the objective function with respect to b yields

$$dL(b) = \sum_{j=1}^{m} \sum_{k=1}^{n} \left[\frac{\partial f(x(b))}{\partial x_k(b)} - \sum_{i=1}^{m} \pi_i(b) \frac{\partial g(x(b))}{\partial x_k(b)} \right] \frac{\partial x_k(b)}{\partial b_j} db_j$$

$$+ \sum_{j=1}^{m} \sum_{i=1}^{m} \frac{\partial \pi_i(b)}{\partial b_j} [g_i(x(b)) - b_i] db_j + \sum_{j=1}^{m} \pi_j(b) db_j. \tag{6.52}$$

By the first-order necessary conditions for optimality, this reduces to

$$dL(b) = \sum_{j=1}^{m} \pi_j(b) db_j. \tag{6.53}$$

Holding the remaining prices constant, we have

$$0 = \pi_i db_i + \pi_j db_j$$
$$-\frac{db_i}{db_j} = \frac{\pi_j}{\pi_i}. \tag{6.54}$$

6.9.   Summary

This chapter introduces several mathematical programming models that have been used to analyze the effect of risk on the farm firm. However, this review is not exhaustive and the treatment of any individual formulation is far from complete. The interested reader is referred to Turvey, Escalante, and Nganje (2005) for additional references. However, this presentation should serve as an introduction to some of the more popular formulations.

Chapter 7

Risk Efficiency
Approaches — Stochastic Dominance

To this point, most of this book follows the premise that individuals choose between actions or alternatives to maximize their expected utility. Axiomatically, this assumption is based on Bernoulli's principle, which was expanded upon by Von Neumann and Morgenstern. While a more complete form of the axioms required for von Neumann and Morgenstern is presented in Chapter 3, the axioms required can be simplified into three in particular:

Ordering and transitivity: A person either prefers one of two risky prospects a_1 and a_2 or is indifferent between them. Further, if the individual prefers a_1 to a_2 and a_2 to a_3, then he prefers a_1 to a_3.

Continuity: If a person prefers a_1 to a_2 to a_3, then there exists some subjective probability level $\Pr[a_1]$ such that he is indifferent between the gamble paying a_1 with probability $\Pr[a_1]$ and a_3 with probability $1 - \Pr[a_1]$, which leaves him indifferent with a_2.

Independence: If a_1 is preferred to a_2, and a_3 is any other risky prospect, a lottery with a_1 and a_3 outcomes will be preferred to a lottery with a_2 and a_3 outcomes when $\Pr[a_1] = \Pr[a_2]$. In other words, preference between a_1 and a_2 is independent of a_3.

However, some literature has raised questions regarding the adequacy of these assumptions: Allais (1953) raised questions about the axiom of independence; May (1954) and Tversky (1969) questioned the transitivity of preferences. These studies question whether preferences under uncertainty are adequately described by the traditional expected utility framework. One alternative to the expected utility hypothesis is to develop risk efficiency criteria rather than expected utility axioms. Risk efficiency criteria are an attempt to reduce the collection of all possible alternatives to a smaller collection of risky alternatives that contain the optimum choice.

One example of risk efficiency criteria was the use of expected value–variance formulations to derivation of optimum portfolios (as presented in Chapter 5). The expected value–variance frontier contains the set of possible portfolios such that no other portfolio can be constructed with a higher return with the same relative risk measured as the variance of the portfolio. The contention is that this efficient set contains the utility maximizing portfolio. Chapter 3 demonstrates under what conditions the expected value–variance framework is consistent with expected utility. Now instead of using expected utility justifying risk efficiency, risk efficiency measures are justified under their own right (i.e., the axiomatic formulations for risk efficiency criteria are developed without relying on expected utility). An alternative justification of risk efficiency measures involves the scenario, where the individual's risk preferences are difficult to elicit.

7.1. Stochastic Dominance

One of the most frequently used *risk efficiency* approaches is stochastic dominance. To demonstrate the concept of stochastic dominance, consider the simplest form (*first-order stochastic dominance*). Assume that the decision maker is faced with two alternative investments, a and b. Further, assume that the probability density function for alternative a can be characterized by the probability density function $f(x)$. Similarly, assume that the return on investment b is associated with the probability density function $g(x)$. Investment a is said to be the first-order dominant of investment b if and only if

$$G(x) = \int_{-\infty}^{x} g(s)ds \geq \int_{-\infty}^{x} f(s)ds = F(x) \qquad (7.1)$$

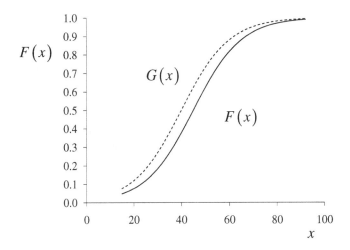

Fig. 7.1. Graphical depiction of first-degree stochastic dominance.

for all x. This result is depicted graphically in Fig. 7.1. Thus, investment a is always more likely to yield a higher return. Intuitively, one investment will dominate the other investment if their cumulative distribution functions do not cross. Economically, the only axiom required for first-degree stochastic dominance (FSD) is that the individual prefers more to less, or is nonsatiated in consumption. This basic criteria would appear noncontroversial, however, it is not very discerning.

7.1.1. *The Concept of an Efficiency Criteria*

An *efficiency criteria* is a decision rule for dividing alternatives into two mutually exclusive groups: efficient and inefficient. If an alternative is in the efficient group, then it is one that an investor may choose. An inefficient investment will not be chosen by any investor regardless of individual risk preferences. From an economic standpoint, the criteria should be related to general notions of utility or preferences. In general, the more global the preference, the less discerning the criteria (i.e., the fewer alternatives eliminated). A smaller efficient set requires more stringent requirements on preferences.

7.1.1.1.   First-Degree Stochastic Dominance

The most general efficiency criteria relies only on the assumption that utility is nondecreasing in income, or the decision maker prefers more of at least one good to less.

FSD Rule: Given two cumulative distribution functions F and G, an option F will be preferred to the second option G by FSD independent of concavity if $F(x) \leq G(x)$ for all return x with at least one strict inequality.

Intuitively, this rule states that one alternative F will dominate G if its cumulative distribution function always lies to the left of G's. Mathematically, FSD can be justified by the integrals of the utility function times each alternative distribution function:

$$E_F U(x) = \int_a^b U(x) f(x) dx \quad \text{versus} \quad E_G U(x) = \int_a^b U(x) g(x) dx.$$

$$(7.2)$$

Note that the utility function is the same for each investment alternative, but the distribution function changes. If investment F dominates investment G, then the difference, Δ, defined as

$$\Delta = E_F U(x) - E_G U(x) = \int_a^b [f(x) - g(x)] U(x) dx \qquad (7.3)$$

must be greater than or equal to zero for all x with at least one strict inequality. Integrating by parts yields

$$\Delta = (U(x)[F(x) - G(x)]|_a^b + \int_a^b [G(x) - F(x)] U'(x) dx. \quad (7.4)$$

The first part of the integral is equal to zero since $F(a), G(a) = 0$ and $F(b), G(b) = 1$. The sign of the Δ is then positive if $U'(x)$ is positive for all x and $G(x) \geq F(x)$ for all x.

7.1.1.2.   Second-Degree Stochastic Dominance

As noted in the preceding section, the only requirement for FSD to be valid is that the decision makers prefer more to less. Building on this concept, *second-degree stochastic dominance* (SSD) invokes risk aversion by inferring that the utility function is concave, implying that the second derivative of the utility function is negative.

SSD Rule: A necessary and sufficient condition for an alternative F to be preferred to a second alternative G by all risk averse decision makers is that

$$\int_{-\infty}^{x} F(z)dz \leq \int_{-\infty}^{x} G(z)dz \tag{7.5}$$

for all x, or

$$\int_{-\infty}^{x} [G(z) - F(z)]dz \geq 0 \tag{7.6}$$

with at least one inequality.

Mathematically, starting from the result from Eq. (7.4)

$$\Delta = \int_{a}^{b} [G(x) - F(x)] U'(x)dx$$

$$= \left(U'(x) \int_{a}^{b} [G(x) - F(x)]dx \Big|_{a}^{b} \right.$$

$$\left. + \int_{a}^{b} \left(-U''(x) \int_{a}^{x} [G(z) - F(z)]dz \right) dx \right., \tag{7.7}$$

which is positive if

$$\int_{a}^{x} [G(z) - F(z)]dz \leq 0 \tag{7.8}$$

for all x with a strict inequality for at least one x.

Graphically, another explanation of SSD can be determined by alternative F dominates alternative G for all risk averse individuals if the cumulative area under F exceeds the area under the cumulative distribution function G for all values x, or if the cumulative area between F and G is nonnegative for all x.

Table 7.1. Observed gross margins.

	Corn	Soybean	Wheat
1	133.9547	117.0221	106.7415
2	160.9626	154.2518	118.1192
3	136.3654	135.6167	102.3978
4	145.1248	133.0063	127.3458
5	134.4932	122.7777	115.3229
6	122.7883	114.4257	108.2281
7	123.0762	121.1467	111.1333
8	100.5129	102.8930	118.9812
9	129.9954	143.0956	114.0733
10	134.2391	134.0585	116.8623

Example: corn, soybean, wheat production

From the test, the observed gross margins per acre for corn, soybeans, and wheat are presented in Table 7.1. Sorting by level of profit, a step cumulative distribution function

$$F(x_n) = \frac{n}{N}, \qquad x_n = \{x_1, x_2, \ldots, x_N\}, \qquad (7.9)$$

where x_n denotes the collection of crop profits sorted in ascending order. Hence, the cumulative distribution function can be constructed for each crop as depicted in Table 7.2.

The results in Table 7.2 indicate that corn does not dominate either soybeans or wheat (or vice versa) because the difference is neither always nonpositive or nonnegative. However, soybeans dominate wheat in the first degree because the difference in the empirical distribution functions is consistently negative. Hence, the cumulative distribution function for soybeans is consistently below the cumulative distribution function for wheat.

7.1.2.   *Increasing Risk*

Next, the notion of stochastic dominance with respect to a function is developed from Meyer (1975, 1977a,b). However, in working through

Table 7.2. Comparing CDFs for each crop.

Income	CDF (corn)	CDF (soybeans)	CDF (wheat)	Corn-soybeans	Corn-wheat	Soybeans-wheat
100.51	0.1	0.0	0.0	0.1	0.1	0.0
102.40	0.1	0.0	0.1	0.1	0.0	−0.1
102.89	0.1	0.1	0.1	0.0	0.0	0.0
106.74	0.1	0.1	0.2	0.0	−0.1	−0.1
108.23	0.1	0.1	0.3	0.0	−0.2	−0.2
111.13	0.1	0.1	0.4	0.0	−0.3	−0.3
114.07	0.1	0.1	0.5	0.0	−0.4	−0.4
114.43	0.1	0.2	0.5	−0.1	−0.4	−0.3
115.32	0.1	0.2	0.6	−0.1	−0.5	−0.4
116.86	0.1	0.2	0.7	−0.1	−0.6	−0.5
117.02	0.1	0.3	0.7	−0.2	−0.6	−0.4
118.12	0.1	0.3	0.8	−0.2	−0.7	−0.5
118.98	0.1	0.3	0.9	−0.2	−0.8	−0.6
121.15	0.1	0.4	0.9	−0.3	−0.8	−0.5
122.78	0.1	0.5	0.9	−0.4	−0.8	−0.4
122.79	0.2	0.5	0.9	−0.3	−0.7	−0.4
123.08	0.3	0.5	0.9	−0.2	−0.6	−0.4
127.35	0.3	0.5	1.0	−0.2	−0.7	−0.5
130.00	0.4	0.5	1.0	−0.1	−0.6	−0.5
133.01	0.4	0.6	1.0	−0.2	−0.6	−0.4
133.95	0.5	0.6	1.0	−0.1	−0.5	−0.4
134.06	0.5	0.7	1.0	−0.2	−0.5	−0.3
134.24	0.6	0.7	1.0	−0.1	−0.4	−0.3
134.49	0.7	0.7	1.0	0.0	−0.3	−0.3
135.62	0.7	0.8	1.0	−0.1	−0.3	−0.2
136.37	0.8	0.8	1.0	0.0	−0.2	−0.2
143.10	0.8	0.9	1.0	−0.1	−0.2	−0.1
145.12	0.9	0.9	1.0	0.0	−0.1	−0.1
154.25	0.9	1.0	1.0	−0.1	−0.1	0.0
160.96	1.0	1.0	1.0	0.0	0.0	0.0

Meyer's articles, concepts from a couple of other important pieces (Diamond and Stiglitz, 1974; Pratt, 1964; Rothschild and Stiglitz, 1970) are required. Meyer (1975) present a definition of *increasing risk* that yields an ordering in terms of riskiness over a larger class of cumulative distributions than the ordering obtained using Rothschild and Stiglitz's original definition.

Assume that x and y are random variables with cumulative distributions F and G, respectively. In general, the goal of this literature is to identify the conditions, where G can be said to be at least as risky as F. Further, the literature attempts to make this ranking among cumulative distributions without numerous assumptions regarding the utility function. That is, F will be preferred to G by an agent with utility function $u(x)$ if

$$\int_0^1 u(x)dF(x) \geq \int_0^1 u(x)dG(x). \tag{7.10}$$

Also assume that the utility function is a continuous, twice differentiable function so that the Arrow–Pratt risk aversion coefficient can be defined.

Rothschild and Stiglitz proposed three definitions of increasing risk to eliminate weaknesses associated with using variance as a measure of risk. Specifically, increasing risk can be defined by (1) $G(x)$ is at least as risky as $F(x)$ if $F(x)$ is preferred or indifferent to $G(x)$ by all risk averse agents, (2) $G(x)$ is at least as risky as $F(x)$ if $G(x)$ can be obtained from $F(x)$ by a sequence of steps that shift weight from the center of $f(x)$ to its tails without changing the mean (this is referred to as mean-preserving spreads), or (3) $G(y)$ is at least as risky as $F(x)$ if y is a random variable that is equal in distribution to x plus some random noise.

Rothschild and Stiglitz found that necessary and sufficient conditions on the cumulative distributions $F(x)$ and $G(x)$ for $G(x)$ should be at least as risky as $F(x)$ are

$$\int_0^1 [G(x) - F(x)]dx \geq 0, \quad \forall y \in [0, 1]$$

$$\int_0^1 [G(x) - F(x)]dx \neq 0. \tag{7.11}$$

Rothschild and Stiglitz show that this definition yields a partial ordering over the set of cumulative distributions in terms of riskiness. The ordering is partial in two senses: (1) only cumulative distributions of the same mean can be ordered and (2) not all distributions with the same mean can be ordered. Thus, a necessary, but not sufficient condition for one distribution to be riskier than another by Rothschild and Stiglitz is that there means be equal.

Diamond and Stiglitz extended Rothschild and Stiglitz defining increasing risk as $G(x)$ is at least as risky as $F(x)$ if $G(x)$ can be obtained from $F(x)$ by a sequence of steps, each shift weight from the center of $f(x)$ to its tails while keeping the expectation of the utility function, $u(x)$, constant.

7.1.2.1.   *Definition Based on Unanimous Preference*

Rothschild and Stiglitz's first definition defines $G(x)$ to be as risky as $F(x)$ if $F(x)$ is preferred or indifferent to $G(x)$ by all risk averse agents. In other words, $F(x)$ is unanimously preferred by the class of agents known as risk averse. Restating the general principle $G(x)$ is at least as risky as $F(x)$ if $F(x)$ is unanimously preferred or indifferent to $G(x)$ by all agents who are at least as risk averse as a risk neutral agent. This restatement shifts the focus from the absolute level of risk aversion to the relative level of risk aversion. This leads to a redefinition:

Definition 7.1. Cumulative distribution $G(x)$ is at least as risky as cumulative distribution $F(x)$ if there exists some agent with a strictly increasing utility function $u(x)$ such that for all agents more risk averse than the original agent $F(x)$ is preferred or indifferent to $G(x)$.

7.1.2.2.   *Mean Preserving Spread*

A *mean preserving spread* is a function which when added to the density function transfers weight from the center of the density function to its tails without changing the mean. Formally, $s(x)$ is a spread if

$$\int_0^1 s(x)dx = 0. \tag{7.12}$$

$s(x)$ changes sign at most twice in $(0, 1)$, $s(x) \geq 0$ for all $x \leq z$ for some $z \in (0, 1)$, and $s(x_0) > 0$ for some $x_0 \in (0, z)$. Thus, the mean-preserving spread moves probability mass to the outside of the probability density function to yield a riskier investment.

Definition 7.2. Cumulative distribution $G(x)$ is at least as risky as cumulative distribution $F(x)$ if $G(x)$ can be obtained from $F(x)$ by a finite sequence of cumulative distributions

$$F_1(x) = s_1(x)F(x),$$
$$F_2(x) = s_2(x)F_1(x), \ldots, G(x) = s_n(x)F_{n-1}(x), \qquad (7.13)$$

where each $F_i(x)$ differs from $F_{i-1}(x)$ by a single spread.

Implications of Definitions

Theorem 7.1. *Consider cumulative distributions $F(x)$ and $G(x)$, then there is an increasing continuous function $r(x)$ such that:*

$$\int_0^y [G(x) - F(x)]dr(x) \geq 0, \quad \forall y \in [0,1] \qquad (7.14)$$

if and only if $G(x)$ is at least as risky as $F(x)$ by Definition 7.1.

Proof.   Assume there exists an increasing twice differentiable $r(x)$ such that

$$\int_0^y [G(x) - F(x)]dr(x) \geq 0, \quad \forall y \in [0,1]. \qquad (7.15)$$

Let $z = r(x)$ and define functions $G^*(.)$ and $F^*(.)$ by $G^*(z) = G(r^{-1}(z))$ and $F^*(z) = F(r^{-1}(z))$. Then,

$$\int_0^y [G^*(r^{-1}(x)) - F^*(r^{-1}(x))]dr(x) \geq 0, \quad \forall y \in [0,1] \qquad (7.16)$$

or

$$\int_{r(0)}^{r(y)} [G^*(z) - F^*(z)]dz \geq 0, \quad \forall y \in [0,1]. \qquad (7.17)$$

By a previous proof (Hadar and Russell, 1969) that

$$\int_{r(0)}^{r(y)} [G^*(z) - F^*(z)]u'(z)dz \geq 0$$

$$\Rightarrow \int_{r(0)}^{r(y)} [G^*(z) - F^*(z)]du(z) \geq 0. \qquad (7.18)$$

$\square$

A couple of notes on this point: Making the assumption that $r(x)$ is a one-to-one mapping, Eq. (7.18) is simply a change in the definition of the cumulative distribution, to defining the integral on the variable z, which is related to the original variable x by $z = r(x)$. This assumption would be implied by the imposition $x = r^{-1}(z)$. Along the same lines, the bounds of integration really have not changed, they have only mapped into the variable z.

Mapping the transformation back to x by $v(x) = u(r(x))$, the integral becomes

$$\int_0^1 [G(x) - F(x)]dv(x) \geq 0. \qquad (7.19)$$

Integrating this relationship by parts yields

$$\int_0^1 v(x)dF(x) \geq \int_0^1 v(x)dG(x) \qquad (7.20)$$

for all such $r(x)$.

Next, using the result from Pratt to show that this inequality holds for all utility functions more risk averse than $r(x)$. Theorem 1 from Pratt (which is used by Meyer) then states that

$$r_1(x) \geq r_2(x)$$

$$\Rightarrow \begin{cases} \pi_1(x, z) \geq \pi_2(x, z) \\ u_1(u_2^{-1}(t)) \quad \text{is strictly convex at } t \\ \dfrac{u_1(y) - u_1(x)}{u_1(w) - u_1(v)} \leq \dfrac{u_2(y) - u_2(x)}{u_2(y) - u_2(v)} \quad \text{for } v < w \leq x < y. \end{cases}$$

$$(7.21)$$

The point of the Pratt theorem is that for any individual more risk averse than $r(x)$ the inequality will also hold. Thus, $F(x)$ will be preferred by all agents who are more risk averse than the index individual.

Theorem 7.2. *Consider cumulative distribution functions $F(x)$ and $G(x)$, such that their probability functions cross a finite number of times, then there is an increasing continuous twice differentiable function $r(x)$, such that*

$$\int_0^y [G(x) - F(x)] dr(x) \geq 0, \quad \forall y \in [0, 1] \tag{7.22}$$

if and only if $G(x)$ is at least as risky as $F(x)$ by Definition 7.2.

Next, Meyer shows that his proposed definition of increasing risk yields a partial ordering over the set of cumulative distributions. In order for the definition to provide a partial ordering, it is necessary to show that the definition is (1) binary, (2) transitive, (3) reflexive, and (4) antisymmetric.

Define the relationship $F \leq_r G$ to mean that G is at least as risky as F. To show transitivity then requires $F \leq_r G$ and $G \leq_r H$ implies that $F \leq_r H$. This proof is completed by defining $U(r(x))$ as the set of all utility functions such that $u(x)$ is more risk averse than $r(x)$. Given this definition, $F \leq_r G$ by all individuals such that $u(x) \subset U(r_1(x))$ and $G \leq_r H$ by all individuals such that $u(x) \subset U(r_2(x))$.

Given these two definitions, derive $r_3(x)$ as the level of risk aversion such that $U(r_3(x)) = U(r_1(x)) \cup U(r_2(x))$. This implies that $r_3(x)$ is that set of individuals who are both more risk averse than $r_1(x)$ and $r_2(x)$. Which implies that

$$-\frac{r_3''(x)}{r_3'(x)} = \max\left(-\frac{r_1''(x)}{r_1'(x)}, -\frac{r_2''(x)}{r_2'(x)}\right). \tag{7.23}$$

Given that expected utility is transitive for economic agents, this result is sufficient to show that individuals who prefer G to H and F to G also prefer F to H.

To show that the dominance relationship is antisymmetric, note that

Theorem 7.3. $F <_r G$ *and* $G <_r F$ *if and only if* $F = G$.

The proof of this theorem is apparent by the integral definition. $F <_r G$ requires that

$$\int_0^y [G(x) - F(x)]dr_1(x) \geq 0, \quad \forall y \in [0, 1]. \tag{7.24}$$

Similarly, $G <_r F$ would require that

$$\int_0^y [F(x) - G(x)]dr_2(x) \geq 0, \quad \forall y \in [0, 1]. \tag{7.25}$$

Again defining $r_3(x)$ as that of risk averse agents who are more risk averse than both $r_1(x)$ and $r_2(x)$ yields

$$-\frac{r_3''(x)}{r_3'(x)} = \max\left(-\frac{r_1''(x)}{r_1'(x)}, -\frac{r_2''(x)}{r_2'(x)}\right). \tag{7.26}$$

Thus, for some group of risk averse agents

$$\int_0^y [G(x) - F(x)]dr_3(x) \geq 0, \quad \forall y \in [0, 1] \tag{7.27}$$

and

$$\int_0^y [G(x) - F(x)]dr_3(x) \geq 0, \quad \forall y \in [0, 1]. \tag{7.28}$$

The only way to meet both of these conditions is for $G = F$ for all x.

Theorem 7.4. *For any two cumulative distributions $F(x)$ and $G(x)$ such that their probability functions cross a finite number of times there exists an increasing continuous twice differentiable function, $r(x)$, such that*

$$\int_0^y [G(x) - F(x)]dr(x) \geq 0, \quad \forall y \in [0, 1]. \tag{7.29}$$

7.1.2.3.   *Risk Aversion with Respect to a Function*

The general idea is to restrict the risk aversion coefficient for stochastic dominance to those risk aversion coefficients in a given interval $r_1(x) < r(x) < r_2(x)$. This problem will be solved by finding the utility function $u(x)$ which satisfies:

$$r_1(x) \leq -\frac{u''(x)}{u'(x)} \leq r_2(x), \quad \forall x \in [0, 1] \tag{7.30}$$

and minimizes

$$\int_0^1 [G(x) - F(x)]u'(x)dx. \tag{7.31}$$

Given that this integral yields the expected value of $F(x)$ minus the expected value of $G(x)$, the minimum will be greater than zero if $F(x)$ is preferred to $G(x)$ by all agents who prefer $F(x)$ to $G(x)$. If the minimum is less than zero, then the preference is not unanimous for all agents whose risk aversion coefficients are in the state range. Another problem is that utility is invariant to a linear transposition. Thus, we must stipulate that $u'(0) = 1$.

The problem is then to use the control variable $-u''(x)/u'(x)$ to maximize the objective function

$$-\int_0^1 [G(x) - F(x)]u'(x)dx. \tag{7.32}$$

subject to the equation of motion

$$(u'(x))' = -u'(x)\frac{u''(x)}{u'(x)} \tag{7.33}$$

and the control constraints

$$-\frac{u''(x)}{u'(x)} - r_1(x) \geq 0$$

$$\frac{u''(x)}{u'(x)} + r_2(x) \geq 0 \tag{7.34}$$

with the initial condition $u'(0) = 1$.

Rewriting the problem, substituting $z(x) = -u''(x)/u'(x)$ yields

$$\max - \int_0^1 [G(x) - F(x)]u'(x)dx$$

$$s.t. \ (u'(x))' = u'(x)z(x) \qquad (7.35)$$

$$-z(x) + r_1(x) \leq 0$$

$$z(x) - r_2(x) \leq 0.$$

The traditional Hamiltonian for this problem then becomes

$$H = -[G(x) - F(x)]u'(x) + \lambda(x)(u'(x)z(x)). \qquad (7.36)$$

Following the basic Pontryagin results, a given path for the control variable is optimum given three conditions:

(1) Optimality condition

$$\frac{\partial H}{\partial z} = 0. \qquad (7.37)$$

(2) Multiplier or Costate condition

$$-\frac{\partial H}{\partial u'} = \lambda'(x). \qquad (7.38)$$

(3) Equation of motion

$$\frac{\partial H}{\partial \lambda} = (u'(x))'. \qquad (7.39)$$

However, in the current scenario the Hamiltonian has to be amended to account for the two inequality constraints. Specifically,

$$L = H + \mu_1(x)(-z(x) + r_1(x)) + \mu_2(x)(z(x) - r_2(x)). \qquad (7.40)$$

First, examine the derivatives of the Lagrangian with respect to $z(x)$, the control variable. Hence,

$$\frac{\partial L}{\partial z(x)} = \lambda(x)u'(x) - \mu_1(x) + \mu_2(x) = 0. \qquad (7.41)$$

Given the restriction that the Lagrange multipliers must be nonnegative at optimum

$$z(x) = \begin{cases} r_1(x) & \text{if } \lambda(x)u'(x) \geq 0 \Rightarrow \mu_1(x) \geq 0 & \text{with } \mu_2(x) = 0 \\ r_2(x) & \text{if } \lambda(x)u'(x) \leq 0 \Rightarrow \mu_2(x) \geq 0 & \text{with } \mu_1(x) = 0 \end{cases}.$$

$$(7.42)$$

Thus, the minimum value of the integral occurs at one of the boundaries. The question is then: What determines which boundary? To examine this begin with the Costate condition:

$$-\frac{\partial L}{\partial u'(x)} = [G(x) - F(x)] - \lambda(x)z(x) = \lambda'(x). \qquad (7.43)$$

Given this expression and working backward from the transversality condition yields the appropriate solution. Specifically, the transversality condition for this control problem specifies that

$$\lambda(1)u'(1) = 0. \qquad (7.44)$$

Given this boundary condition, the value of $\lambda(x)u'(x)$ can be defined by the integral

$$\int_x^1 \frac{d(\lambda(s)u'(s))}{ds} ds = \lambda(1)u'(1) - \lambda(x)u'(x)$$

$$= -\lambda(x)u'(x) \qquad (7.45)$$

$$\lambda(x)u'(x) = -\int_x^1 \frac{d(\lambda(s)u'(s))}{ds} ds.$$

Hence, to complete the derivation, the value of the derivative under the integral must be determined

$$\frac{d(\lambda(x)u'(x))}{dx} = \lambda'(x)u'(x) + \lambda(x)u''(x). \qquad (7.46)$$

Substituting for $\lambda'(x)$ from the Costate condition yields

$$\frac{d(\lambda(x)u'(x))}{dx} = ([G(x) - F(x)] - \lambda(x)z(x))u'(x) + \lambda(x)u''(x).$$

$$(7.47)$$

Substituting for $z(x)$ and canceling like terms yields

$$\frac{d(\lambda(x)u'(x))}{dx} = [G(x) - F(x)]u'(x). \tag{7.48}$$

Yielding the expression

$$\lambda(x)u'(x) = -\int_x^1 [G(x) - F(x)]u'(x)dx. \tag{7.49}$$

Merging this result with the previous derivations yields the infamous Theorem 7.5.

Theorem 7.5. *An optimal control* $-u''(x)/u'(x)$ *that maximizes*

$$-\int_0^1 [G(x) - F(x)]u'(x)dx$$

$$s.t. \; r_1(x) \le -\frac{u''(x)}{u'(x)} \le r_2(x) \tag{7.50}$$

$$u'(0) = 1$$

is given by

$$-\frac{u''(x)}{u'(x)} = \begin{cases} r_1(x) & \text{if } \int_x^1 [G(x) - F(x)]u'(x)dx \le 0 \\[2ex] r_2(x) & \text{if } \int_x^1 [G(x) - F(x)]u'(x)dx \ge 0 \end{cases}. \tag{7.51}$$

The empirical idea is then to determine whether one distribution is dominated by another distribution in the second degree with respect to a particular set of preferences. To do this, the question is if the integral switches from negative to positive within the range of risk aversion coefficients. Theorem 7.5 states that the risk aversion that maximizes the integral will be found at one boundary or the other depending on the sign of the integral.

Application: Assume that $G(x) - F(x)$ is always nonnegative. Then

$$\int_y^1 [G(x) - F(x)]u'(x) \ge 0, \quad \forall y \in [0, 1] \tag{7.52}$$

for any $u'(x)$ considered. Thus, the optimal control for this particular $F(x)$ and $G(x)$ is to choose $-u''(x)/u'(x)$ equal to its maximum possible; i.e., $-u''(x)/u'(x) = r_2(x)$ at all points x. Similarly, if $F(x)$ and $G(x)$ are such that $G(x) - F(x)$ is always nonpositive, then an optimal choice is to set $-u''(x)/u'(x)$ equal to its minimum throughout.

If $G(x) - F(x)$ changes sign a finite number of times, then it is known that for some x_0, $G(x) - F(x)$ does not change sign in the interval $[x_0, 1]$. Thus, the optimal solution in $[x_0, 1]$ is given by Eq. (7.51), and once the solution in $[x_0, 1]$ is known, the solution of $[0, x_0]$ can be calculated by Theorem 7.5.

7.2. Applications of Stochastic Dominance

The risk of a firm can be decomposed into business risk and financial risk. Business risk is the risk inherent in the production/business environment. Financial risk results from the leverage (debt) decisions based on that business environment. Collins (1985) demonstrated this decomposition using a DuPont expansion:

$$R_E = \frac{\left(\frac{r_p}{A} + i\right) - \delta K}{1 - \delta},$$
(7.53)

where R_E is the rate of return to equity, r_p is the return to production activities, A is the total level of output, i is the rate of capital gains, δ is the debt-to-asset position, and K is the cost of debt capital. Following this development,

$$R_A = \frac{r_p}{A} + i \Rightarrow \sigma_A^2.$$
(7.54)

Therefore, the expected rate of return on equity and the variance of the rate of return on equity can be derived as

$$\mu_E = \frac{\mu_A - \delta K}{1 - \delta}$$

$$\sigma_E^2 = \frac{\sigma_A^2}{(1 - \delta)^2}$$
(7.55)

assuming that the cost of debt is nonstochastic.

The linkage between debt and asset returns has been addressed several ways in economic literature. Following Modigliani–Miller, the value of assets is independent of the capital structure. The Modigliani–Miller results assume infinite arbitrage opportunities. Debt can be increased by issuing bonds and using the proceeds to buy stock (increasing the relative level of debt). Alternatively, someone could sell stock and buy bonds to increase the relative level of equity.

In agriculture, such arbitrage is typically not possible (i.e., there are very few publicly traded firms). Literature following Collins allows for endogeneity, but holds the cost of debt constant. Moss, Ford, and Castejon (1991) allows the cost of capital to be stochastic:

$$R_t^i = P_t^i Y_t - V - Dr_t, \qquad (7.56)$$

where R_t^i is the return on investment i in period t, P_t^i is the price of output using marketing instrument i in period t, Y_t is the yield in period t, V is the variable cost, D is the level of debt, and r_t is the cost of debt capital in period t.

Moss, Ford, and Castejon then applies the above definition to derive a series of returns for oranges marketing their output as FCOJ using cash markets, futures, and participation. Applying second-degree stochastic dominance, Moss, Ford, and Castejon derived the efficient choices of marketing instruments based on the solvency level presented in Table 7.3. Given that the choice of marketing instrument does not change within the each marketing period, what is the role of risk aversion? To see this, focus on the February marketing period. The study then finds that the choice of marketing instrument only changes between debt-to-asset positions when risk aversion is taken into account. A similar study is Gloy and Baker (2002).

A second application is taken from Moss and Livanis (2009). This study focuses on the implementation of the stochastic dominance approach. Following the standard first- and second-degree stochastic dominance approach, dominance is defined as

$$\Delta = E_F U(x) - E_G U(x) = \int_a^b [f(x) - g(x)] U(x) dx \geq 0. \qquad (7.57)$$

Table 7.3.   Second-degree stochastic dominance results.

Solvency ratio	Participation	Cash	Hedge
December			
0	X		
30	X		
40	X		
50	X		
60	X		
February			
0	X	X	
30	X	X	
40	X	X	
50	X	X	
60	X	X	
April			
0		X	
30		X	
40		X	
50		X	
60		X	

Empirically, following the example in Table 7.2 first-degree stochastic dominance can be determined by the expression

$$\tilde{\Delta}_1 = G(x) - F(x) \geq 0 \quad \forall x. \qquad (7.58)$$

To rigorously apply this rule,

$$\tilde{\Delta}_1^i = \inf_x G(x) - F(x)$$

$$\tilde{\Delta}_1^s = \sup_x G(x) - F(x). \qquad (7.59)$$

Recalling the results from Table 7.2, one distribution dominates the other distribution as long as $\tilde{\Delta}_1^i$ and $\tilde{\Delta}_1^s$ have the same sign ($\tilde{\Delta}_1^i \tilde{\Delta}_1^s \geq 0$).

Table 7.4. Second-degree stochastic dominance with respect to a function results.

Solvency ratio	Participation	Cash	Hedge
$r(x) = [0.000002, 0.000003]$			
0	X	X	
30		X	
40		X	
50		X	
60		X	
$r(x) = [0.000003, 0.000004]$			
0	X		
30	X	X	
40	X	X	
50		X	
60		X	
$r(x) = [0.000004, 0.000005]$			
0	X		
30	X		
40	X		
50	X	X	
60	X	X	

Extending these results to second-degree stochastic dominance, Eq. (7.7) can be expressed as

$$\Delta_2 = \left(U'(x) \int_a^b [G(x) - F(x)]dx \Big|_a^b \right.$$
$$\left. + \int_a^b \left(-U''(x) \int_a^x [G(z) - F(z)]dz \right) dx \right). \quad (7.60)$$

Assuming that neither distribution in Eq. (7.60) is dominated in the first degree, the rule for second-degree stochastic dominance (Table 7.4) becomes

$$\tilde{\Delta}_2 = \int_a^x [G(z) - F(z)]dz \geq 0 \quad \forall x. \quad (7.61)$$

Again, to empirically apply the dominance relationship, define

$$\tilde{\Delta}_2^i = \inf_x \int_a^x [G(z) - F(z)] dz$$

$$\tilde{\Delta}_2^s = \sup_x \int_a^z [G(z) - F(z)] dz. \tag{7.62}$$

As in the application of first-degree stochastic dominance, if $\tilde{\Delta}_2^i$ and $\tilde{\Delta}_2^s$ have the same sign, or $\tilde{\Delta}_2^i \tilde{\Delta}_2^s \geq 0$, one distribution dominates the other.

The difficulty lies in the empirical form of the distribution function. Specifically, the empirical distribution function is typically defined as

$$F(x) = \frac{N^*[y \leq x]}{N}, \tag{7.63}$$

which is a step function (a continuous form of Eq. (7.9)). Moss and Livanis propose fitting a continuous form of Eq. (7.63) using a nonparametric regression. The general nonparametric regression or kernel regression can be expressed as

$$\hat{y}_i = \sum_{j=1}^N k(x_j, \hat{x}_i) y_j$$

$$k(x_j, \hat{x}_i) = \frac{1}{\sqrt{2\pi}} \exp[-\delta(x_j - \hat{x}_i)], \tag{7.64}$$

where $k(.)$ is the kernel weighting function that weights observations based on the distance between the sample point x_j, δ is the bandwidth that controls how the weight on each point declines as the data moves away from the point of approximation and the point of estimation $\hat{x}_i$. This weighted average gives the estimated value of the dependent variable $\hat{y}_i$ without parametric restrictions (except for the weighting function and the bandwidth). In this case, the dependent variable is $F(x)$ or the step cumulative distribution function from Eq. (7.63). Rewriting the kernel function to account for the discrete data as

$$k^*(x_j, \hat{x}_i) = \frac{k(x_j, \hat{x}_i)}{\sum_{l=1}^N k(x_l, \hat{x}_i)} \tag{7.65}$$

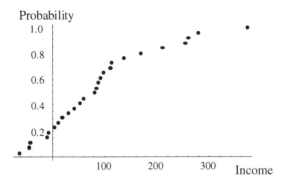

Fig. 7.2. Empirical cumulative distribution function from Moss and Livanis.

the height of the cumulative distribution function at any point x_k can be written as

$$\tilde{F}(x_k) = \frac{k(x_j, x_k)}{\sum_{l=1}^{N} k(x_l, x_k)} F(x_j). \qquad (7.66)$$

Figure 7.2 depicts the step distribution function computed using Eq. (7.4) for the data in Moss and Livanis. To see the possible implications of the nonparametric formulation, Fig. 7.3 presents the nonparametric probability density function implied by taking the

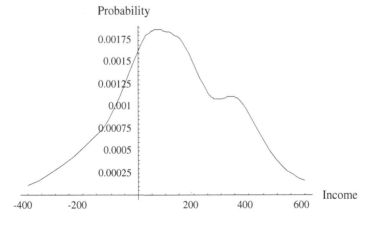

Fig. 7.3. Probability density function for cotton using nonparametric regression.

derivative of Eq. (7.66) with respect to x_k. Note that this distribution is clearly nonnormal. Specifically, it exhibits the beginning of a second mode.

Given the general form of the cumulative distribution function in Eq. (7.66), the empirical difficulty involves computing the supremum and infimum in Eq. (7.59). The difference function between the two cumulative density functions is not smooth enough to admit standard numerical search algorithms such as a Newton–Raphson. Hence, more complex search algorithms are required. Note that all that is required to demonstrate that one distribution does not dominate another is two points with different signs. Hence, Moss and Livanis apply a random search to generate the point of divergence. However, nonderivative based algorithms such as genetic algorithms may provide good procedures to ascertain dominance.

7.3.   Summary

While the axiomatic basis of the expected utility theory is sound, several studies have raised empirical difficulties. This chapter describes the application of risk efficiency criteria implemented through stochastic dominance. The chapter presents the most general forms of stochastic dominance (first- and second-degree stochastic dominance) and how these concepts can be linked to expected utility. In addition, the chapter describes how these basic forms can be implemented. Next, a more general formulation that allows for the introduction of risk aversion is developed. Finally, the chapter ends with a couple of examples.

Chapter 8

Dynamic Decision Rules and the Value of Information

The previous chapters have been largely silent about the timing aspects of risk. Implicitly, the risk had two time periods. In the first time period, a decision had to be made (i.e., whether to invest in a given opportunity) and in the second time period, the result became apparent. This chapter expands the scenario to three or more periods. Specifically, a decision is made in the first period, in the second period a random event is observed, which may or may not contain information about the eventual outcome, and in the third period the result is realized. The question then involves the ability to change the decision in the second period. This framework is extended in Chapter 10 with the introduction of options. However, this chapter focuses on a more rudimentary form of the decision which allows for the development of the value of information.

8.1. Decision Making and Bayesian Probabilities

Traditionally, Bayesian analysis involves a procedure whereby new information is integrated into a prior distribution to generate an updated or posterior distribution. At times the concept of Bayesian probability theory is confused with the subjective probability theory where an individual has an intuition regarding the probability of an event, instead of a frequency view of probability theory which strives for an objective version of probability.

At the base of Bayesian inference is Bayes' equation:

$$P[a|b] = \frac{P[a,b]}{P[b]}, \qquad (8.1)$$

where $P[a \mid b]$ is the probability of the event a occurring such that event b has already occurred, $P[a, b]$ is the joint probability of both event a and event b occurring, and $P[b]$ is the marginal probability of event b occurring (or the probability of event b occurring such that event a has been integrated from the distribution). For a detailed discussion of Bayesian statistics, see Section 2.3 of Chapter 2. Again, this chapter is interested in the change in information over time and how this change in information can be used in decision making under risk.

This basic concept of *prior* and *posterior probabilities* can be used to develop one manifestation of the value of information. Specifically, imagine the state space depicted in Fig. 8.1. This diagram depicts the potential outcomes of two random events, each of which has two potential outcomes. The first event yields outcomes O_1 and O_2 each of which occurs with probability 0.5. The second event results in $O_{1|1}$ and $O_{2|1}$ if O_1 occurred in the first event and $O_{1|2}$ and $O_{2|2}$ given that O_2 occurred in the first event. Intuitively, picture $O_{1|1}$ and $O_{1|2}$ as the same event with a different intervening event. Similarly, $O_{2|1}$ and $O_{2|2}$ are the same event. What does change with the intervening event is the relative probability that each outcome will occur. Given O_1, the probability of outcome 1 in the second stage ($O_{1|1}$) is 0.7 compared

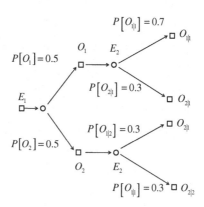

Fig. 8.1.   State-space for the value of information problem.

with a probability of outcome 2 in the second stage ($O_{1|2}$) of 0.3. The difference in the probability of the payoffs given the outcome in the first stage gives rise to the value of information. Next, consider alternatives. First, assume that alternative A_1 pays \$10 in state 1 and \$0 in state 2. Second, assume alternative A_2 pays \$5 in state 1 and \$4 in state 2. To determine the expected value of the investment, first determine the probabilities that state 1 and state 2 will occur. For states 1 and 2, the total probabilities are

$$P[S_1] = P[O_1]P[O_{1|1}] + P[O_2]P[O_{1|2}] = 0.5(0.7) + 0.5(0.3) = 0.5$$
$$P[S_2] = P[O_1]P[O_{2|1}] + P[O_2]P[O_{2|2}] = 0.5(0.3) + 0.5(0.7) = 0.5.$$
$$(8.2)$$

The expected value of each alternative is then:

$$E[A_1] = P[S_1]V[A_1 \mid S_1] + P[S_2]V[A_1 \mid S_2] = 0.5(10) + 0.5(0) = \$5.00$$
$$E[A_2] = P[S_1]V[A_2 \mid S_1] + P[S_2]V[A_2 \mid S_2] = 0.5(5) + 0.5(4) = \$4.50.$$
$$(8.3)$$

Thus, in the absence of risk aversion, the decision maker would choose A_1 with an expected value of \$5.00.

The next question is: What is the initial signal worth? Starting from the last node and working backward, assume that O_1 has occurred, what is the optimum decision?

$$
\begin{aligned}
E[A_1 \mid O_1] &= P[O_{1|1}]V[A_1 \mid S_1] + P[O_{2|1}]V[A_1 \mid S_2] \\
&= 0.7(10) + 0.3(0) = \$7.00 \\
E[A_2 \mid O_1] &= P[O_{1|1}]V[A_2 \mid S_1] + P[O_{2|1}]V[A_2 \mid S_2] \\
&= 0.7(5) + 0.3(4) = \$4.70.
\end{aligned}
\qquad (8.4)
$$

Thus, just like the scenario without the intervening event, choose A_1. However, the result is somewhat different given that O_2 occurs. Specifically,

$$
\begin{aligned}
E[A_1 \mid O_2] &= P[O_{1|2}]V[A_1 \mid S_1] + P[O_{2|2}]V[A_1 \mid S_2] \\
&= 0.3(10) + 0.7(0) = \$3.00 \\
E[A_2 \mid O_2] &= P[O_{1|2}]V[A_2 \mid S_1] + P[O_{2|2}]V[A_2 \mid S_2] \\
&= 0.3(5) + 0.7(4) = \$4.30.
\end{aligned}
\qquad (8.5)
$$

Under this scenario, the decision maker would choose A_2 over A_1. The decision rule is then to choose A_1 if event O_1 occurs and action A_2 if event O_2 occurs. The expected value of this strategy is:

$$V[Strategy] = P[O_1]E[A_1 \mid O_1] + P[O_2]E[A_2 \mid O_2]$$
$$= 0.5(7.00) + 0.5(4.30) = \$5.65. \tag{8.6}$$

The value of the information is then the difference between the conditioned and unconditioned decision:

$$V[Information] = 5.65 - 5.00 = \$0.65. \tag{8.7}$$

Hence, if the investor is risk neutral, he is willing to pay up to $0.65 for the information in event E_2.

Bringing in risk aversion, assume utility is negative exponential with a risk aversion coefficient of $\rho = 0.01$. Compute the optimum strategy without observing E_2

$$E[U(A_1)] = 0.5[-\exp(-\rho10)] + 0.5[-\exp(-\rho0)] = -0.9524$$
$$E[U(A_2)] = 0.5[-\exp(-\rho5)] + 0.5[-\exp(-\rho4)] = -0.9560 \tag{8.8}$$

so that A_1 yields a higher level of expected utility than A_2. Using the expenditure function formulation based on negative exponential preferences yields a certainty equivalent of $4.88.

Next, to derive the optimal strategy after observing E_2, compute the expected utility of each value of E_2. The equivalent to Eq. (8.4) under risk aversion assuming that O_1 has occurred implies

$$E[U(A_1) \mid O_1] = 0.7[-\exp(-\rho10)] + 0.3[-\exp(-\rho0)] = -0.9339$$
$$E[U(A_2) \mid O_1] = 0.7[-\exp(-\rho5)] + 0.3[-\exp(-\rho4)] = -0.9541. \tag{8.9}$$

Hence, if O_1 is observed the optimum response is A_1 that yields a certainty equivalent of $6.89. Next, assume that O_2 occurred then

$$E[U(A_1) \mid O_1] = 0.3[-\exp(-\rho10)] + 0.7[-\exp(-\rho0)] = -0.9715$$
$$E[U(A_2) \mid O_1] = 0.3[-\exp(-\rho5)] + 0.7[-\exp(-\rho4)] = -0.9579, \tag{8.10}$$

which implies that the optimal action is A_2 yielding a certainty equivalent of $4.30. Thus, like the risk neutral scenario, the optimal decision changes after the signal is observed. To derive the value of the signal under risk aversion, compute the expected utility given that the optimal rule has been applied

$$V[Strategy] = 0.5E[U(A_1) \mid O_1] + 0.5E[U(A_2) \mid O_2] = -0.9457,$$
(8.11)

which is higher than the larger expected utility in Eq. (8.8). Computing the certainty equivalent of the result in Eq. (8.11) yields $5.59. Hence, the value of the signal under risk aversion is $0.71. Note that under this combination of assumptions the value of the information increases with risk aversion.

This simple scenario points to an important point about dynamic decision rules in general and the value of a signal in particular. Not only must the posterior probability change with the signal, but the optimum decision must change. The first assumption (that the posterior probability changes) is a statistical question while the second assumption is an artifact of the decision problem.

8.2.   Concepts of Information

One way to define information content is to focus on the *entropy* of the signal. The difference in information content then answers the first question in the proceeding section (i.e., whether the posterior distribution is quantitatively different). Following Shannon and Weaver (1963), the entropy of a signal is defined as

$$H = -\sum_{i=1}^{n} p_i \ln(p_i).$$
(8.12)

Intuitively, as H increases the value of the signal decreases. Further, H decreases as one event becomes increasing likely. (In the limit, as the probability of one-event approaches one, the natural log of one goes to zero and the value of the entropy measure approaches zero.)

From Fig. 8.1, the unconditioned probability has equally likely events. The value of the information in that distribution function is

Table 8.1.   Entropy of equally likely events.

State	Probability	$-p_i \ln(p_i)$
S_1	0.5	0.3466
S_2	0.5	0.3466

$H = 0.6931$

Note that $\exp(H) = 2$

Table 8.2.   Entropy of unequally likely events.

State	Probability	$-p_i \ln(p_i)$
S_1	0.7	0.2497
S_2	0.3	0.3712

$H = 0.6109$

Note that $\exp(H) = 1.84202$

depicted in Table 8.1 while the posterior probability given event O_1 occurs is depicted in Table 8.2. Thus, equally likely events contain 0.6931 entropy units of information while the 70/30 outcome has 0.6109 entropy units of information. From a statistical perspective, the 50/50 event contains more entropy.

The next level of complexity is added by the concepts of prior and posterior probabilities. Assume that there are two signals, p and q, each with two potential outcomes. One question is: What is the value of the information in signal q given the information already observed in signal p? Theil (1967) gives the value of such as signal as:

$$I(p, q) = \sum_{i=1}^{n} p_i \ln\left(\frac{p_i}{q_i}\right). \tag{8.13}$$

Table 8.3 depicts the information content of the conditional structure using the information inequality in Eq. (8.13). Following the results in the preceding section, information in the informative prior case in Table 8.3 could yield a different decision while the information in the

Table 8.3. Information in prior signal.

State	Probability p	Probability q	$p_i \ln(p_i/q_i)$
Informative Prior			
S_1	0.5	0.7	-0.1682
S_2	0.5	0.3	0.2554
			$I = 0.0872$
Uninformative Prior			
S_1	0.5	0.5	0.0000
S_2	0.5	0.5	0.0000
			$I = 0.0000$

uninformative prior would not yield a change in decisions. Empirically, note that the information value from Eq. (8.13) increases as the signal becomes informative. Hence, the quantity of information in the signal is an indication of the potential value of the forecast. However, without ancillary information on whether the decision changes as the information measure increases, the entropy measure cannot measure the value of information in a signal.

Another statistically based definition of information comes from the estimation process via the likelihood function (as discussed in Chapter 2). The estimation process involves choosing the values of parameters that maximize the likelihood of a particular sample. Under normality, the natural log of the likelihood for a linear regression can be written as

$$L = -\frac{T}{2} \ln \left(\frac{\left[\sum_{t-1}^{T} \{ y_t - \alpha_0 - \alpha_1 x_t \} \right]^2}{T} \right) + \text{const.} \qquad (8.14)$$

Finding the parameters α_0 and α_1 that maximize the likelihood function then yields an estimate of these parameters. As discussed in Chapter 2, to find the maximum value of the likelihood function take the derivative of the function with respect to each parameter and set them equal. The matrix of second moments is referred to as the information matrix.

This matrix is the basis for statistical inference and yields such things as the standard error of the parameter estimates.

Again, neither the entropy formulation in Eq. (8.13) nor the likelihood function in Eq. (8.14) directly contains information on changes in tastes or preferences, or whether the changes in probability are sufficient to alter perceptions about random events.

8.3.   A Model of Information

Return to the continuous version of the model in Fig. 8.1 as presented by Chavas and Pope (1984) where the decision maker chooses the level of two inputs x_1 and x_2. Assume that economic agents possess a concave utility function $U(y, x_1, x_2, e_1, e_2)$. The decision maker chooses the level of x_1 and x_2 that maximizes that level of utility

$$\max_{x_1, x_2} \ E_1 U(y, x_1, x_2, e_1, e_2), \qquad (8.15)$$

where y is the initial level of income and e_1 and e_2 are random variables. The notation E_1 denotes expectations taken in the first period in time. The problem as specified in Eq. (8.15) corresponds to the "open loop" solution where no learning takes place (which can be compared with the discrete decisions presented in Eqs. (8.3) and (8.8)).

Given that e_1 can be observed before decisions on x_2 have to be made (as assumed in Eqs. (8.4) and (8.5) in the case of risk neutrality and Eqs. (8.9) and (8.10) under risk aversion), the second phase of the decision process can be rewritten as

$$V(x_1, y, e_1) = \max_{x_2} \ E_2 U(y, x_1, x_2, e_1, e_2), \qquad (8.16)$$

where $V(x_1, y, e_1)$ is the maximum utility in the second period given that the event e_1 is known. From Eq. (8.9), the optimum choice is A_1 if event O_1 occurred. In other words, the right-hand side of Eq. (8.16) implies the optimum value of x_2 conditioned on the new information, or e_1. The real difference between Eqs. (8.16) and (8.9) is the possibility of active learning. Specifically, the only choice in Eq. (8.9) was whether to observe the signal or not. Equation (8.16) allows the level of x_1 to affect the signal observed. If the level of x_1 does not affect the probability of e_2, the system corresponds to passive learning. Hence, the question becomes whether the individual expends resources on x_1 to reduce the uncertainty of the future.

Once, the optimizing behavior in Eq. (8.16) has been determined, x_1 can then be selected to maximize the expectation of this value function:

$$\max_{x_1} E_1 V(x_1, y, e_1). \tag{8.17}$$

It can be shown that

$$E_1 \max_{x_1} V(x_1, y, e_1) \geq \max_{x_1} E_1 V(x_1, y, e_1) \tag{8.18}$$

(Chavas and Pope, 1984, p. 707).

Next, consider the situation where the decision maker must pay for the information in the signal. Assume that the decision maker must pay B_1 to observe the signal. The question is whether the information is "worth" observing? Define the maximum bid price for B_1 as the bid price which equates the two

$$E_1 \max_{x_1} V(x_1, y - B_1, e_1) = \max_{x_1} E_1 V(x_1, y, e_1). \tag{8.19}$$

Chavas and Pope conclude that as long as the marginal utility of income $(\partial V / \partial y > 0)$ is positive, the maximum bid for information is also positive. One implication of this finding is that as long as the marginal utility of income is positive, costless information can never make the decision maker worse off. Taking a Taylor series expansion around B_1 yields the value of the information as

$$B_1 = -\frac{1}{2} \frac{\left[\frac{\partial^2 V}{\partial e_1 \partial x_1} \left(\frac{\partial^2 V}{\partial x_1^2} \right)^{-1} \frac{\partial^2 V}{\partial x_1 \partial e_1} \right]}{\frac{\partial V}{\partial y}}. \tag{8.20}$$

Given that the marginal utility of income is positive and that the utility function is negative semi-definite in x_1, the value of information is positive. Integrating the variance of the signal

$$
B_1 \cong \left\{ \frac{-\frac{1}{2}\left[\frac{\partial^2 V}{\partial e_1 \partial x_1} \left(\frac{\partial^2 V}{\partial x_1^2} \right)^{-1} \frac{\partial^2 V}{\partial x_1 \partial e_1} \right]}{\frac{\partial V}{\partial y}} \right\} \mathrm{Var}(e_1). \qquad (8.21)
$$

Chavas and Pope conclude that information can be considered as an intermediate good in the production process. Hence, the decision maker could choose either to not invest in information (i.e., benefit from passive learning) or to invest in the intermediate output and pursue active learning.

8.4.   Summary

This chapter develops two related economic models where new information is made available to the decision maker in a risky environment. In both scenarios, the basic problem is to attempt to value the information. Two conditions must be met for the information to be valuable. First, following Theil's formulation the information must change the entropy. Second, this change in statistical information must potentially change the decision being made.

Chapter 9

Market Models of Decision Making under Risk

A basic assumption in microeconomic or price theory is that market prices contain all the relevant information about the scarcity of goods. The firm decides the appropriate quantity of goods to supply to the market based on input and output prices. The supply curve provides a conceptual framework for these choices. Similarly, the demand curve provides a conceptual framework for the decisions made by consumers. In this conceptual model, consumers allocate income between available commodities to maximize their well-being. Aggregating this stylized representation across all commodities, the Arrow–Debreau framework suggests that the interaction between supply and demand maximizes overall societal welfare or surplus. The market price then contains all the relevant information about the tradeoffs that society must make in determining the resource allocations. On a similar level, market models of risk explain how the market prices of various investment opportunities are affected by relative risk. Two popular market models for pricing risk are the capital asset pricing model (CAPM) developed by Lintner (1965a,b), Mossin (1966), and Sharpe (1964) and the arbitrage pricing theory (APT) developed by Ross (1976). These formulations assume that the market prices of investments manifest information about each asset's relative risk. However, to support this conclusion, these formulations assume that capital markets are perfect. Specifically, they assume that (1) all goods and investment markets are competitive, (2) all goods and investments are infinitely divisible, (3) information is costless, (4) there are no transaction costs, and (5) no individual is large enough to effect the price.

Starting with the firms, assume that (1) all goods are produced by firms, (2) these firms purchase the factors of production in the first period, produce output, and market their goods in the second period, and (3) these firms do not have any capital of their own and must raise this capital by issuing stock.

From the consumption side assume that consumers begin with an endowment w_1. The consumer's choice problem is then twofold. First, the consumer must decide how much to consume in this period, c_1, and how much to invest, h_1 (this choice is depicted in Fig. 9.1). The investment will earn a return of $h_1(1 + R_i)$ that will be consumed in the second period, c_2. Second, the consumer must decide how to invest h_1, that is, how to divide it up between a wide array of assets.

The consumer's problem implies two decision dimensions. The first is *intertemporal* (across time). In this decision, the consumer has a time preference, the preference between consuming now and consuming later. The second is the risk or uncertainty on the investment. Both of these questions can be represented in the utility function. The market model assumes that firms supply investment and consumers demand

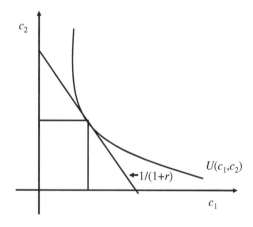

Fig. 9.1.   Consumer's tradeoff between consumption now and consumption later.

investment opportunities. In addition, the formulation is based on the assumption that there exists an equilibrium where the supply of stocks equals the demand of stocks based on their expected returns and relative variance.

9.1.   Risk Equilibrium from the Consumer's Point of View

At the outset, assume that consumers are risk averse and that risk can be characterized using the normal distribution function. Given these assumptions, the expected value–variance formulation can be used, or in this case the expected value-standard deviation approach, to model utility/risk efficiency. Starting as in Chapter 5 by defining the expected return on the portfolio as

$$\mu_p = \sum_{i=1}^{N} \mu_i \tilde{z}_i \qquad (9.1)$$

where μ_p is the expected return from the portfolio, μ_i is the expected return on a specific asset, and $\tilde{z}_i$ is the level of asset i held in a specific portfolio. The standard deviation of the return on the portfolio is then written as

$$\sigma_p = \left[\sum_{j=1}^{N} \sum_{i=1}^{N} \sigma_{ij} \tilde{z}_i \tilde{z}_j \right]^{\frac{1}{2}}, \qquad (9.2)$$

where σ_p is the standard deviation of a particular portfolio and σ_{ij} is the covariance between asset i and asset j. In addition, the portfolio balance condition is imposed so that

$$\sum_{i=1}^{N} \tilde{z}_i = 1. \qquad (9.3)$$

As a starting point for the market model, reformulate the standard deviation of the portfolio to derive the contribution of each asset to the overall risk of the portfolio

$$\sigma_p = \frac{\sigma_p^2}{\sigma_p} = \frac{\sum_{j=1}^{N} \sum_{i=1}^{N} \sigma_{ij} \tilde{z}_i \tilde{z}_j}{\sigma_p}$$

$$= \sum_{j=1}^{N} \tilde{z}_j \left(\frac{\sum_{i=1}^{N} \sigma_{ij} \tilde{z}_i}{\sigma_p} \right). \tag{9.4}$$

The risk of a particular asset z_j is then dependent on weighted covariances between asset j and the returns in the rest of the portfolio. Remember that the z_is are weights in the general portfolio.

This raises two points. First note that the risk of an individual asset depends on the portfolio weights and the risk of the portfolio. Second, the risk of a particular asset depends both on its own variance and the variance of the remaining assets in the portfolio. Thus, as the number of assets becomes large and the portfolio becomes well diversified, the risk of a particular asset is more dependent on the covariance with other assets in the portfolio than on its own risk.

Following the discussion of the expected value–variance frontier in Chapter 5, assume that consumers choose the portfolio that minimizes risk for any given level of expected income. In this case, risk is parametrized by the standard deviation instead of the variance.

$$\min \sigma_p = \left(\sum_{j=1}^{N} \sum_{i=1}^{N} \sigma_{ij} \tilde{z}_j \tilde{z}_i \right)^{\frac{1}{2}}$$

$$s.t. \, \mu_p = \sum_{i=1}^{N} \mu_i \tilde{z}_i = \mu^* \tag{9.5}$$

$$\sum_{i=1}^{N} \tilde{z}_i = 1.$$

Forming the Lagrangian to solve the constrained optimization problem

$$L = \sigma_p + \lambda_1 \left(\mu_p - \sum_{i=1}^{N} \mu_i \tilde{z}_i \right) + \lambda_2 \left(1 - \sum_{i=1}^{N} \tilde{z}_i \right)$$

$$\frac{\partial L}{\partial \tilde{z}_k} = \frac{\partial \sigma_p}{\partial \tilde{z}_k} - \lambda_1 \mu_k - \lambda_2 = 0$$

$$\frac{\partial L}{\partial \lambda_1} = \mu_p - \sum_{i=1}^{N} \mu_i \tilde{z}_i = 0$$

$$\frac{\partial L}{\partial \lambda_2} = 1 - \sum_{i=1}^{N} \tilde{z}_i = 0, \tag{9.6}$$

where λ_1 is the Lagrange multiplier for the expected rate of return constraint and λ_2 is the Lagrange multiplier for the portfolio balance constraint. Each of the Lagrange multipliers has important implications. λ_1 will become the inverse of the slope of the capital market line (CML). Specifically, λ_1 is the change in standard deviation (e.g., risk) associated with a change in the required expected return. Thus, $1/\lambda_1$ is the change in expected return associated with a change in risk (or the slope of the line tangent to the efficient expected value–standard deviation frontier presented in Fig. 9.2). Solving the first-order conditions with respect to $\tilde{z}_k$ by equating the value of λ_2 for any two investments k and l yields

$$\frac{\partial \sigma_p}{\partial \tilde{z}_k} - \lambda_1 \mu_k = \frac{\partial \sigma_p}{\partial \tilde{z}_l} - \lambda_1 \mu_l$$

$$\mu_k - \mu_l = \frac{1}{\lambda_1} \left(\frac{\partial \sigma_p}{\partial \tilde{z}_k} - \frac{\partial \sigma_p}{\partial \tilde{z}_l} \right). \tag{9.7}$$

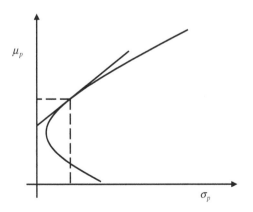

Fig. 9.2. Tangency between the capital market line and the efficient portfolio.

Summing this result over all stocks yields

$$\sum_{i=1}^{N} \tilde{z}_i \left(\mu_j - \mu_i \right) = \frac{1}{\lambda_1} \left(\sum_{k=1}^{N} \tilde{z}_k \frac{\partial \sigma_p}{\partial \tilde{z}_j} - \sum_{k=1}^{N} \tilde{z}_k \frac{\partial \sigma_p}{\partial \tilde{z}_i} \right) \qquad (9.8)$$

this expression is the expected return of selling investment i to buy investment j, or the expected value of the portfolio that is long in investment j and short in investment i. Focusing on the short sale of the market portfolio, let i in Eq. (9.8) become the market portfolio. This relationship becomes

$$\mu_j - \mu_p = \frac{1}{\lambda_1} \left(\sum_{k=1}^{N} \tilde{z}_k \frac{\partial \sigma_p}{\partial \tilde{z}_j} - \frac{\partial \sigma_p}{\partial \tilde{z}_p} \right) \qquad (9.9)$$

that is $\sum_{i=1}^{N} \tilde{z}_i \mu_j$ and $\sum_{i=1}^{N} \tilde{z}_i \mu_i = \mu_p$. Also note that $\sum_{k=1}^{N} \tilde{z}_k (\partial \sigma_p / \partial \tilde{z}_k) = \partial \sigma_p / \partial \tilde{z}_p$. The standard deviation of the market portfolio can be expressed as

$$\sigma_p = \sum_{j=1}^{N} \tilde{z}_j^* \left(\frac{\sum_{i=1}^{N} \sigma_{ij} \tilde{z}_i^*}{\sigma_p} \right) \Leftarrow \frac{\partial \sigma_p}{\partial \tilde{z}_k} = \frac{\sum_{i=1}^{N} \sigma_{ik} \tilde{z}_i}{\sigma_p}$$

$$\sum_{j=1}^{N} \tilde{z}_j^* \frac{\partial \sigma_p}{\partial \tilde{z}_j} = \sum_{j=1}^{N} \tilde{z}_j^* \sum_{i=1}^{N} \frac{\sigma_{ij} \tilde{z}_i^*}{\sigma_p}$$

$$= \sum_{j=1}^{N} \sum_{i=1}^{N} \frac{\sigma_{ij} \tilde{z}_i^* \tilde{z}_j^*}{\sigma_p} \qquad (9.10)$$

$$= \sigma_p,$$

where $\tilde{z}_i^*$ are the relative shares of the market portfolio. Substituting the result of Eq. (9.10) into Eq. (9.9) and given that $\partial \sigma_p / \partial \tilde{z}_k = \sum_{i=1}^{N} \sigma_{ik} / \sigma_p$

$$\mu_j - \mu_p = \frac{1}{\lambda_1} \left(\sum_{i=1}^{N} \frac{\sigma_{ij} \tilde{z}_i \tilde{z}_j}{\sigma_p} - \sigma_p \right)$$

$$= \frac{1}{\lambda_1} \left(\frac{\sigma_{jp}}{\sigma_p} - \sigma_p \right). \qquad (9.11)$$

The last relationship in Eq. (9.11) gives an implicit form of the line that is tangent to the expected value–standard deviation frontier depicted in Fig. 9.2 for any efficient expected value–standard deviation combination (Sharpe, 1964).

9.2. The Role of the Riskless Asset

The equilibrium presented above does not yield an estimable representation because different investors may have different risk preferences. One way around this ambiguity is to introduce a riskless asset. The riskless asset reduces the potential number of efficient portfolios to a single portfolio. Within this equilibrium, there is only one efficient portfolio of assets. Any degree of risk aversion can construct a risk preferred position by holding a combination of the single efficient portfolio and borrowing or lending at the riskless rate. Substituting $\tilde{z}_m$ for $\tilde{z}_p$ or letting the index portfolio be the market efficient portfolio yields

$$\mu_j - \mu_m = \frac{1}{\lambda_1} \left(\frac{\partial \sigma_m}{\partial \tilde{z}_j} - \sigma_m \right). \tag{9.12}$$

Substituting the risk-free asset for μ_j and given that the covariance between the risk-free asset return and the return on the market portfolio is zero yields

$$\frac{1}{\lambda_1} = \frac{(\mu_m - r_f)}{\sigma_m}. \tag{9.13}$$

This relationship solves for the tangency between the expected value–standard deviation frontier such that the capital market line originates from the risk-free asset (Fig. 9.3). Specifically, Eq. (9.13) determines the slope of the expected value–standard deviation frontier such that the implied return equals the risk-free rate of return when the standard deviation becomes zero. Building on this relationship, from Eq. (9.7) yields

$$\mu_k - \mu_l = \frac{1}{\lambda_1} \left(\frac{\partial \sigma_p}{\partial \tilde{z}_l} - \frac{\partial \sigma_p}{\partial \tilde{z}_k} \right) \ni : k \to j, l \to f$$

$$\Rightarrow \mu_j - r_f = \frac{1}{\lambda_1} \frac{\partial \sigma_p}{\partial \tilde{z}_j} \tag{9.14}$$

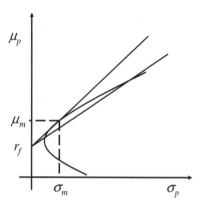

Fig. 9.3. Tangency between capital market line and the expected value–standard deviation frontier.

or let portfolio k become a single asset j and let asset l become the risk-free asset (implying $\partial \sigma_p / \partial \tilde{z}_k = 0$). Replacing the expression $1/\lambda_1$ in Eq. (9.14) with the result from Eq. (9.13) yields

$$\mu_j = r_f + \left[\frac{\mu_m - r_f}{\sigma_m} \right] \frac{\partial \sigma_m}{\partial \tilde{z}_j}$$

$$= r_f + \left[\frac{\mu_m - r_f}{\sigma_m} \right] \frac{\sigma_{jm}}{\sigma_m}$$

$$= r_f + \beta_j (\mu_m - r_f), \tag{9.15}$$

where

$$\beta_j = \frac{\sigma_{jm}}{\sigma_m^2}. \tag{9.16}$$

Eq. (9.15) can be respecified as the CAPM regression equation

$$r_{it} = \alpha_i + \beta_i r_{mt} + \varepsilon_{it}. \tag{9.17}$$

9.3.   Risk Equilibrium from the Firm's Perspective

The consumer's problem is then to first determine the optimal portfolio of risky assets (i.e., the tangency between the expected value–standard deviation frontier associated with the risk-free asset) and then to determine the combination of borrowing and lending (i.e., current consumption and future consumption) that maximizes expected utility. This equilibrium then provides the demand for equity capital. The other scissor of the market equilibrium is then the supply of investment opportunities provided by firms seeking capital. In the capital market model, this supply is typically in the form of equity capital investments (stocks).

9.3.1. *Deriving the Security Market Line*

In capital market equilibrium, the expected value–standard deviation frontier depicted in Fig. 9.3 and Eq. (9.14) equals the security market line (SML) depicted in Fig. 9.4, which is consistent with the standard CAPM relationship

$$E[r_j] = \bar{r} + \beta[E(r_m) - \bar{r}]. \tag{9.18}$$

Starting with a two-asset portfolio, construct a portfolio using investment i and asset m.

$$
\begin{aligned}
E[R_p] &= aE[R_i] + (1-a)E[R_m] \\
\sigma(R_p) &= [a^2\sigma_i^2 + 2(1-a)a\sigma_{im} + (1-a)^2\sigma_m^2]^{\frac{1}{2}}.
\end{aligned}
\tag{9.19}
$$

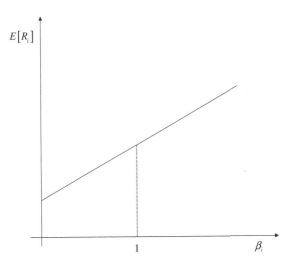

Fig. 9.4. The security market line.

Next, examine the risk–return relationship based on changes in the share of asset i.

$$\frac{\partial E[R_p]}{\partial a} = E[R_i] - E[R_m]$$

$$\frac{\partial \sigma(R_p)}{\partial a} = \frac{1}{2}\left[a^2\sigma_i^2 + 2(1-a)a\sigma_{im} + (1-a)^2\sigma_m^2\right]^{-\frac{1}{2}} \quad (9.20)$$

$$\times \left[2a\sigma_i^2 - 2\sigma_m^2 + 2a\sigma_m^2 + 2\sigma_{im} - 4a\sigma_{im}\right].$$

Consider what happens as the share held in asset i becomes small

$$\frac{\partial \sigma(R_p)}{\partial a}\bigg|_{a \to 0} = \frac{\sigma_{im} - \sigma_m^2}{\sigma_m}$$

$$= \left[\frac{\sigma_{im} - \sigma_m^2}{\sigma_m}\right]\frac{\sigma_m}{\sigma_m} = [\beta - 1]\sigma_m. \quad (9.21)$$

The risk–return relationship as the share in asset i becomes small, is then derived as

$$\therefore \frac{\partial E\frac{[R_p]}{\partial a}}{\partial \sigma\frac{(R_p)}{\partial a}}\bigg|_{a \to 0} = \frac{E[R_i] - E[R_m]}{[\beta - 1]\sigma_m}$$

$$= \frac{1}{\beta - 1}\frac{E[R_i] - E[R_m]}{\sigma_m}. \quad (9.22)$$

This relationship then yields

$$\frac{E[R_m] - r_f}{\sigma_m}$$

$$\therefore \frac{1}{\beta - 1}\frac{E[R_i] - E[R_m]}{\sigma_m} = \frac{E[R_m] - r_f}{\sigma_m}$$

$$E[R_i] - E[R_m] = (\beta - 1)[E[R_m] - r_f] \quad (9.23)$$

$$= \beta[E[R_m] - r_f] - E[R_m] + r_f$$

$$\Rightarrow E[R_i] - r_f = \beta[E[R_m] - r_f].$$

The result in Eq. (9.23) implies that the premium (or increased expected return over the risk-free interest rate) is directly proportional to the premium on the market portfolio and the relative riskiness of the individual asset measured by its "beta" (β).

9.3.2.　*Supply of Stocks from the Firm*

To complete the market model for stocks, start by assuming that each firm sells stocks at a price P_i. Investors are willing to bid on these stocks based on the future value of the firm at the end of the year, V_i. The bid price and the value at the end of the year determine the rate of return:

$$R_i = \frac{V_i - P_i}{P_i}. \qquad (9.24)$$

Given that the future value of the firm implies some risk, the rate of return is risky. In addition, given the preceding proof, investors value the investment under the capital market equilibrium. Mathematically, the price and value of the market portfolio becomes

$$P_m = \sum_{i=1}^{N} \tilde{z}_i P_i \quad \text{and} \quad V_m = \sum_{i=1}^{N} \tilde{z}_i V_i, \qquad (9.25)$$

where P_m is the price of the portfolio and V_m is the value of total portfolio (or the value of the portfolio at the end of the holding period). Given these values, define the return on the portfolio (R_m) as

$$V_m = P_m(1 + R_m). \qquad (9.26)$$

Using the result of Eq. (9.13), the aggregate return on the portfolio then implies that

$$\frac{E(V_j) - P_j}{P_j} = R_f + \left[\frac{E(R_m) - R_f}{\sigma_m}\right] \frac{\sigma_{jm}}{P_j \sigma_m}. \qquad (9.27)$$

Thus, assuming that the expected value of the portfolio at the end of the holding period is exogenous along with the covariance between the portfolio return and the return on the market portfolio,

Eq. (9.21) can be used to solve for the market clearing price of the portfolio (P_j).

9.3.3.   *Tests of the CAPM*

The typical estimation procedure for empirically testing the CAPM is a two-step model. First, the annual returns are estimated as a function of the returns on the market portfolio:

$$R_{jt} = a_j + b_j R_{mt}. \tag{9.28}$$

This estimation gives the estimate of the individual beta parameters. It is anticipated that $b_j > 0$ for all stocks indicating a positive price for risk. Using the estimated results from the first estimation, the SML is estimated across equations

$$\bar{R}_j = \hat{\gamma}_0 + \hat{\gamma}_1 b_j + u_j. \tag{9.29}$$

This regression estimates whether the betas are consistent across equations.

The "testable" implications of the CAPM are (1) the intercept term $\hat{\gamma}_0$ should be equal to zero; (2) beta should be the only factor that explains the rate of return on a risky asset. If other terms such as the residual variance, dividend yield, price/earnings ratio, firm size, or beta squared are significant, then the market equilibrium does not hold; (3) the relationship in beta should be linear; (4) the coefficient $\hat{\gamma}_1$ should be equal to $R_{mt} - R_{ft}$; (5) when the equation is estimated over long periods, the rate of return on the market portfolio should be greater than the risk free rate of return.

9.3.4.   *Incorporating Risk using CAPM*

First, the results of the CAPM can be used to make investment decisions by computing the risk adjusted discount rate (RADR). Starting with a single-period return

$$\tilde{R}_j = \frac{\tilde{P}_e - P_0}{P_0}. \tag{9.30}$$

From the CAPM model

$$E[R_j] = R_f + [E[R_m] - R_f]\beta = \frac{E[\tilde{P}_e] - P_0}{P_0}$$

$$R_f + [E[R_m] - R_f]\beta = \frac{E[P_e]}{P_0} - 1$$

$$1 + R_f + [E[R_m] - R_f]\beta = \frac{E[P_e]}{P_0} \tag{9.31}$$

$$P_0 = \frac{E[P_e]}{1 + R_f + [E[R_m] - R_f]}.$$

Alternatively, the results of CAPM can be used to make investment decisions using the certainty equivalent approach

$$1 + R_f + [E[R_m] - R_f]\beta = \frac{E[\tilde{P}_e]}{P_0}$$

$$1 + R_f = \frac{E[\tilde{P}_e] - [E[R_m] - R_f]\beta P_0}{P_0} \tag{9.32}$$

$$P_0 = \frac{E[\tilde{P}_e] - [E[R_m] - R_f]\beta P_0}{1 + R_f}.$$

Both risk adjustments come directly from Eq. (9.18). Reducing the preceding expression to the CAPM formula

$$R_j = R_f + \beta_j(R_m - R_f). \tag{9.33}$$

This risk-adjusted discount rate can be used in present value analysis

$$PV = \sum_{i=1}^{N} \frac{E[CF \mid I_t]}{(1 + [R_f + \beta_j\{R_m - R_f\}])^i}. \tag{9.34}$$

Reformulating this equality slightly

$$\frac{1}{\beta_j}(R_j - R_f) = R_M - R_f. \tag{9.35}$$

This approach leads to transforming the annual rates of return into certainty equivalents (note that this definition is different than the use of certainty equivalent in other parts of this book) based on the market portfolio:

$$PV = \sum_{i=1}^{N} \frac{\left(\frac{1}{\beta_j}\right) E[CF \mid I_t]}{(1 + R_m)^i}. \tag{9.36}$$

Hence, the CAPM model affords a way to incorporate riskiness into firm-level decisions. Moss, Weldon, and Muraro (1991) use this approach to develop different discount rates for investment in Florida citrus. However, it must be remembered that each approach (RADR or certainty equivalents) results in very different answers. The RADR compounds the risk over time through changing the discount rate while the certainty equivalent approach implies a constant effect of risk.

9.4. Arbitrage Pricing Theorem

The arbitrage pricing theory (APT) is an alternative formulation of a market-clearing model for risk. It is somewhat different in that it proceeds from a collection of axioms (or assumptions) about efficient capital markets. Specifically, the concept is to show under what conditions a riskless, wealthless trade provides no expected rate of return. The basic construct for this argument is the arbitrage portfolio $\tilde{z}$ (or a collection of short and long positions for a collection of stocks). The arbitrage portfolio is defined as that set of purchases and sales that leaves wealth unchanged

$$\tilde{z}'1 = 0. \tag{9.37}$$

Instead of using the expected value–variance formulation as a starting point, the APT assumes that investment returns can be written as a *factor model*

$$R_t = \bar{E} + bF_t + \varepsilon_t, \tag{9.38}$$

where $\underset{\sim}{R}$ is the vector of returns on assets ($R \in M_{n \times 1}$), $\bar{E}$ is the vector of expected returns on the assets ($\bar{E} \in M_{n \times 1}$), b is a matrix of factor loadings relating changes in the common factors with the fluctuations in asset returns ($b \in M_{n \times k}$), and ε is a vector of idiosyncratic risks ($\varepsilon \in M_{n \times 1}$).

The typical assumptions regarding the factor model are

$$E[\varepsilon_t \varepsilon'_t] = \theta \ \{\text{a diagonal matrix}\}$$

$$E[\varepsilon_t \varepsilon'_{t-s}] = 0$$

$$E[\varepsilon_t F'_t] = 0 \tag{9.39}$$

$$E[F_t F'_t] = \Psi \ \{\text{a diagonal matrix}\}.$$

Thus, there is no correlation between idiosyncratic risk and factors. The effect of the arbitrage portfolio is then:

$$\tilde{z}' R_t = \tilde{z}' \bar{E} + \tilde{z}' b F_t + \tilde{z}' \varepsilon_t. \tag{9.40}$$

In scalar form

$$\sum_i z_i R_{it} = \sum_i z_i \bar{E}_i + \sum_i z_i b_{i1} F_{1t} + \cdots + \sum_i z_i b_{i2} F_{kt} + \sum_i z_i \varepsilon_{it}. \tag{9.41}$$

The expected return on any portfolio given $E[F_{jt}] = E[\varepsilon_t] = 0$ is

$$E[\tilde{z}' R_t] = \tilde{z}' \bar{E}$$

$$\sum_i z_i R_{it} = \sum_i z_i \bar{E}_i, \tag{9.42}$$

which must be zero for a profitless arbitrage from Eq. (9.37).

The condition is that the trade must be riskless. Taking the variance of the factor model.

$$E[\tilde{z}' R R' \tilde{z}] - [E[\tilde{z}' R]][E[\tilde{z}' R]]'$$

$$= E([\tilde{z}' \bar{E} + \tilde{z}' b F + \tilde{z}' \varepsilon][\bar{E}' \tilde{z} + F' b' \tilde{z} + \varepsilon' \tilde{z}]) - (\tilde{z}' \bar{E} \bar{E}' \tilde{z})$$

$$= E(\tilde{z}' \bar{E} \bar{E}' \tilde{z} + \tilde{z}' \bar{E} F' b' \tilde{z} + \tilde{z}' \bar{E} \varepsilon' \tilde{z} + \tilde{z}' b F \bar{E}' \tilde{z} + \tilde{z}' b F F' b' \tilde{z}$$

$$+ \tilde{z}' b F \varepsilon' \tilde{z} + \tilde{z}' \varepsilon \bar{E}' \tilde{z} + \tilde{z}' \varepsilon F' b' \tilde{z} + \tilde{z}' \varepsilon \varepsilon' \tilde{z}) - (\tilde{z}' \bar{E} \bar{E}' \tilde{z}) \tag{9.43}$$

since $E[F] = 0$, $E[\varepsilon] = 0$, $E[F\varepsilon] = 0$, $E[FF'] = \Psi$, and $E[\varepsilon \varepsilon'] = \theta$

$$V[\tilde{z}' R] = \tilde{z}' b \Psi b' \tilde{z} + \tilde{z}' \theta \tilde{z}. \tag{9.44}$$

The final condition is therefore

$$\tilde{z}'b = 0. \tag{9.45}$$

This actually means that

$$\sum_i \tilde{z}_i b_{i1} = 0$$

$$\sum_i \tilde{z}_i b_{i2} = 0 \tag{9.46}$$

$$\vdots$$

$$\sum_i \tilde{z}_i b_{ik} = 0.$$

The three conditions that underlie the APT are thus: (1) $\tilde{z}'1 = 0 \Rightarrow \sum_i z_i = 0$ or no change in wealth, (2) $\tilde{z}'\bar{E} = 0 \Rightarrow \sum_i z_i \bar{E}_i = 0$ or no profit, and (3) $\tilde{z}'b = 0$ no risk.

The algebraic consequence of this statement is that

$$\bar{E}_i = \lambda_0 + \lambda_1 b_{i1} + \cdots + \lambda_k b_{ik} \tag{9.47}$$

or the expected return on an asset is a linear function of the factor loadings. This result is identical to the CAPM results if there is a single factor.

9.4.1. *Single-Factor Model*

Abstracting away from the specific form of the CAPM model, a single-factor model can be posited as

$$r_i = a_i + \sum_{k=1}^{K} b_{ik} \tilde{f}_k + \tilde{\varepsilon}_i. \tag{9.48}$$

In this model, the random return on an investment r_i is a linear function of some random factor $\tilde{f}_i$ and an idiosyncratic term $\tilde{\varepsilon}_i$.

This factor specification implies

$$E(\tilde{\varepsilon}_i) = E(\tilde{f}_k) = E(\tilde{\varepsilon}_i \tilde{\varepsilon}_j) = E(\tilde{\varepsilon}_i \tilde{f}_k) = E(\tilde{f}_k \tilde{f}_l) = 0$$

$$E(\tilde{\varepsilon}_i^2) = s_i^2 < S_i^2 \tag{9.49}$$

$$E(\tilde{f}_k^2) = 1.$$

Assuming that the *idiosyncratic risk* goes to zero (or that the portfolio is large)

$$r_i = a_i + b_i \tilde{f}_i. \tag{9.50}$$

If the b_is of two assets are the same, then the a_is must be the same for an arbitrage-free model. Suppose the researcher is interested in forming a portfolio of two assets with different b_is, $b_i \neq b_j$, $b_i \neq 0$, $b_j \neq 0$

$$
\begin{aligned}
r &= \tilde{z}(a_i + b_i f) + (1 - \tilde{z})(a_j + b_j f) \\
&= \tilde{z} a_i + \tilde{z} b_i f + a_j - \tilde{z} a_j + b_j f - \tilde{z} b_j f \\
&= [\tilde{z}(a_i - a_j) + a_j] + [\tilde{z}(b_i - b_j) + b_j] f. \tag{9.51}
\end{aligned}
$$

Computing the mean and variance of this portfolio yields

$$
\begin{aligned}
E[r] &= \tilde{z}(a_i - a_j) + a_j \\
V[r] &= E\{[\tilde{z}(a_i - a_j) + a_j] + [\tilde{z}(b_i - b_j) + b_j] f\}^2 \\
&\quad - \{\tilde{z}(a_i - a_j) + a_j\}^2 \\
&= E\{\tilde{z}(a_i - a_j) + a_j\}^2 \tag{9.52} \\
&\quad + E\{[\tilde{z}(a_i - a_j) + a_j][\tilde{z}(b_i - b_j) + b_j] f\} \\
&\quad + E\{[\tilde{z}(b_i - b_j) + b_j] f^2\} - \{\tilde{z}(a_i - a_j) + a_j\}^2 \\
&= [\tilde{z}(b_i - b_j) + b_j]^2.
\end{aligned}
$$

Holding the variance of the portfolio equal to zero (or assuming that trading asset i for asset j yields no additional risk) yields

$$
\begin{aligned}
[\tilde{z}(b_i - b_j) + b_j]^2 &= 0 \\
\tilde{z}(b_i - b_j) + b_j &= 0 \tag{9.53} \\
\tilde{z}^* &= \frac{b_j}{b_j - b_i}.
\end{aligned}
$$

Imposing the *zero-variance portfolio*

$$r = \left[\frac{b_j}{b_j - b_i}\right](a_i - a_j) + a_j + \left[\left(\frac{b_j}{b_j - b_i}\right)(b_i - b_j) + b_j\right]f$$

$$= \frac{b_j(a_i - a_j)}{b_j - b_i} + a_j = R, \tag{9.54}$$

where R is the risk-free rate of return. Thus,

$$\frac{(a_i - a_j)}{(b_j - b_i)} = \frac{R - a_j}{b_j} = \frac{R - a_i}{b_i}. \tag{9.55}$$

Substituting $\bar{r}_i = a_i$

$$\frac{R - \bar{r}_i}{b_j} = \frac{(a_i - a_j)}{(b_j - b_i)}$$

$$R - \bar{r}_i = \frac{(a_i - a_j)}{(b_j - b_i)} b_j \tag{9.56}$$

$$R - \frac{(a_i - a_j)}{(b_j - b_i)} b_j = \bar{r}_i.$$

Thus, by substitution

$$\bar{r}_i = R + \frac{(a_i - a_j)}{(b_j - b_i)} b_j \Rightarrow \lambda_1 = \frac{(a_i - a_j)}{(b_j - b_i)} \tag{9.57}$$

$$\bar{r}_i = R + \lambda_1 b_i,$$

where λ_1 is the factor risk premium. In general

$$\bar{r}_i = \lambda_0 + \lambda_1 b_i, \tag{9.58}$$

where λ_0 is the empirical estimate of the risk-free rate of return.

9.4.2.   *Two-Factor Model*

In the case of multi-factor models, suppose that asset returns are generated by a two-factor linear model

$$r_i = a_i + b_i \tilde{f}_1 + c_i \tilde{f}_2. \tag{9.59}$$

A portfolio of these assets then yields

$$\sum_i \tilde{z}_i r_i = \sum_i \tilde{z}_i a_i + \sum_i \tilde{z}_i b_i \tilde{f}_1 + \sum_i \tilde{z}_i c_i \tilde{f}_2. \qquad (9.60)$$

Again to minimize systematic risk

$$\sum_i \tilde{z}_i b_i = \sum_i \tilde{z}_i c_i = 0. \qquad (9.61)$$

If the portfolio is riskless, then it yields zero profit

$$R = \sum_i \tilde{z}_i a_i$$
$$\sum_i \tilde{z}_i a_i - R = 0. \qquad (9.62)$$

Given $\sum_i \tilde{z}_i = 1 \Rightarrow \sum_i \tilde{z}_i R = R$

$$\therefore \sum_i \tilde{z}_i(a_i - R) = 0 \Rightarrow \begin{pmatrix} a_1 - R & a_2 - R & a_3 - R \\ b_1 & b_2 & b_3 \\ c_1 & c_2 & c_3 \end{pmatrix} \begin{pmatrix} \tilde{z}_1 \\ \tilde{z}_2 \\ \tilde{z}_3 \end{pmatrix}$$

$$= \begin{pmatrix} 0 \\ 0 \\ 0 \end{pmatrix} \qquad (9.63)$$

the matrix

$$\begin{pmatrix} a_1 - R & a_2 - R & a_3 - R \\ b_1 & b_2 & b_3 \\ c_1 & c_2 & c_3 \end{pmatrix} \qquad (9.64)$$

must be singular, or the first row (R_1) must be a linear combination of the last rows $(R_2$ and $R_3)$

$$\begin{pmatrix} a_1 - R & a_2 - R & a_3 - R \\ b_1 & b_2 & b_3 \\ c_1 & c_2 & c_3 \end{pmatrix} \Rightarrow R_1 + \lambda_1 R_2 + \lambda_2 R_3 = 0. \qquad (9.65)$$

Therefore,

$$\bar{r}_i - R = a_i - R = \lambda_1 b_i + \lambda_2 c_i \; \forall i$$
$$\text{or}$$
$$\bar{r}_i = \lambda_0 + \lambda_1 b_i + \lambda_2 c_i \; \forall i$$

(9.66)

or the average returns must be a linear function of the two common factors.

9.5.   Empirical Applications of Capital Market Models

Both the CAPM and APT yield a variety of empirical specifications. At one level, the CAPM could be estimated for modifications of the present value formulation (as depicted in Moss, Weldon, and Muraro, 1991). However, most of the interest in these formulations has been to test for the efficiency of capital markets. Recalling the introduction of this chapter, if capital markets are competitive and efficient, then stock prices contain all of the relevant information about the relative riskiness of investment.

9.5.1.   *Capital Asset Pricing Models*

A basic question that must be addressed in the application of both CAPM and APT models is whether a risk-free asset exists (and whether that risk-free asset is constant over time). In the basic Sharpe–Lintner CAPM model

$$E[R_i] = R_f + \beta_{im}(E[R_m] - R_f)$$

$$\beta_{im} = \frac{\text{Cov}[R_i, R_m]}{\text{Var}[R_m]}$$

(9.67)

with minimum algebra

$$E[R_i] - R_f = \beta_{im}(E[R_m] - R_f)$$

(9.68)

redefining

$$\left. \begin{array}{l} Z_i = E[R_i] - R_f \\ Z_m = E[R_m] - R_f \end{array} \right\} \Rightarrow Z_i = \alpha_i + \beta_i Z_m + \varepsilon_i.$$

(9.69)

Table 9.1.   Sharpe–Lintner results.

	11499	12140	15553	18542	19166	19749	21135
α_i	0.0065	−0.0027	0.0077	0.0022	0.0009	−0.0047	0.0021
	(0.0146)	(0.0052)	(0.0032)	(0.0049)	(0.0054)	(0.0109)	(0.0051)
β_i	1.0388	1.3945	0.3051	0.9985	1.0837	−0.0081	0.9326
	(0.3164)	(0.1132)	(0.0703)	(0.1071)	(0.1169)	(0.2366)	(0.1112)

Constructing a dataset of 43 stocks from the Center for Research into Security Prices (CRSP) dataset, using the return on the Standard and Poors 500 portfolio and using the 3-month treasury bill as the market portfolio, Table 9.1 presents the estimated values of the α_is and β_is

An alternative to the model presented by Sharpe and Lintner is the zero-beta model suggested by Black (1972):

$$E[R_i] = E[R_{0m}] + \beta_{im}(E[R_m] - E[R_{0m}]), \qquad (9.70)$$

where R_{0m} is the return on the zero-beta portfolio, or the minimum variance portfolio that is uncorrelated with the market portfolio. However, the model can be estimated assuming that the zero-beta return is unobserved as

$$E[R_i] = \alpha_{im} + \beta_{im}E[R_m], \qquad (9.71)$$

which yields the empirical model.

$$R_{it} = \alpha_i + \beta_i R_{mt} + \varepsilon_{it}. \qquad (9.72)$$

Table 9.2 presents the estimated α_is and β_is using Black's formulation. Table 9.3 offers a side-by-side comparison of the β_is for each stock. Note that the results are remarkably similar. The similarity has

Table 9.2.   Zero-beta portfolio results.

	11499	12140	15553	18542	19166	19749	21135
α_i	0.0064	−0.0042	0.0103	0.0024	0.0006	−0.0006	0.0024
	(0.0149)	(0.0053)	(0.0033)	(0.0050)	(0.0055)	(0.0111)	(0.0052)
β_i	1.0380	1.3910	0.3089	0.9875	1.0861	−0.0094	0.9298
	(0.3159)	(0.1130)	(0.0703)	(0.1069)	(0.1167)	(0.2362)	(0.1110)

Table 9.3.   Comparison of betas.

	Sharpe–Lintner	Black
11499	1.03884	1.03803
12140	1.39450	1.39097
15553	0.30513	0.30890
18542	0.99853	0.98751
19166	1.08365	1.08609
19749	−0.00807	−0.00944

two implications; first, the use of the treasury bills as a risk-free asset consistent with the data, and second, even in cases where a risk-free asset may not exist (as under inflation) the CAPM is still valid.

9.5.2.   Tests for CAPM Efficiency

The forgoing discussion demonstrates how the αs and βs for the CAPM can be estimated. Of course the question becomes: So what? Other than using the coefficients to adjust a net present value analysis, these results could be used to consider whether the capital markets are efficient (or competitive). As a starting point, consider a general formulation of the Sharpe–Lintner problem

$$\hat{\alpha} = \hat{\mu} - \hat{\beta}\hat{\mu}_m, \quad \hat{\mu} = \frac{1}{T}\sum_{t=1}^{T} R_t, \quad \hat{\mu}_m = \frac{1}{T}\sum_{t=1}^{T} Z_{mt}$$

$$\hat{\beta} = \frac{\sum_{t=1}^{T}(Z_t - \hat{\mu})(Z_{mt} - \hat{\mu}_m)}{\sum_{t=1}^{T}(Z_{mt} - \hat{\mu}_m)^2} \tag{9.73}$$

$$\hat{\Sigma} = \frac{1}{T}\sum_{t=1}^{T}(Z_t - \hat{\alpha} - \hat{\beta}Z_{mt})(Z_t - \hat{\alpha} - \hat{\beta}Z_{mt}).$$

Based on this formulation, the distribution of the parameter of the Sharpe–Lintner formulation can be derived as

$$\hat{\alpha} \sim N\left(\alpha, \frac{1}{T}\left[1 + \frac{\hat{\mu}_m^2}{\sigma_m^2}\right]\Sigma\right)$$

$$\hat{\beta} \sim N\left(\beta, \frac{1}{T}\left[\frac{1}{\sigma_m^2}\right]\Sigma\right). \tag{9.74}$$

The test of the efficient capital market can then be formulated as

$$H_0 : \alpha = 0$$

$$H_1 : \alpha \neq 0 \qquad (9.75)$$

$$J_0 = T \left[1 + \frac{\hat{\mu}_m^2}{\sigma_m^2} \right]^{-1} \hat{\alpha}' \Sigma^{-1} \hat{\alpha}.$$

Specifically, if the markets are in equilibrium the basic CAPM implies excess returns (expected returns not explained by the market return) is zero ($\hat{\alpha} = 0$).

9.5.3. Cross-Section Regression

Using either set of betas, the question then becomes whether the expected returns are consistent with their betas. Taking the Sharpe–Lintner model as a basis

$$E[Z_i] = \gamma_0 + \gamma_1 \beta_i. \qquad (9.76)$$

The estimates for Eq. (9.76) are presented in Table 9.4. Two tests: in the Sharpe–Lintner model

$$\gamma_0 = 0 \Rightarrow \frac{0.006062}{0.000592} = 10.23986. \qquad (9.77)$$

The betas explain cross-section variations in expected returns

$$\gamma_1 = 1 \Rightarrow \frac{1.0 - 0.906548}{0.092863} = 1.006343. \qquad (9.78)$$

A little reformulation:

$$E[Z_i] = \gamma_0 + \gamma_1 \beta_i + \gamma_2 D_i, \qquad (9.79)$$

Table 9.4. Cross-sectional regression.

Parameter	Estimate
γ_0	0.006062
	(0.000592)
γ_1	0.906548
	(0.092863)

Table 9.5. Cross-sectional with risk premium for agribusiness.

Parameter	Estimate
γ_0	0.006083
	(0.000605)
γ_1	0.913067
	(0.097456)
γ_2	−0.000350
	(0.001374)

where D_i is a dummy variable, which is equal to one if the stock is an agribusiness stock and zero otherwise. This approach tests for the presence of a risk premium for agribusiness firms. The results for this formulation presented in Table 9.5 suggest that the hypothesis that agribusiness stocks contain a nonsystematic returns (or risk) can be rejected at any conventional level of significance.

9.5.4. *Arbitrage Pricing Model*

An empirical alternative to the CAPM formulation is the arbitrage pricing model (APM), which is the empirical model for the APT. The returns in the arbitrage pricing model are assumed to be determined by a linear factor model

$$R_t = a + bf_t + \varepsilon_t, \quad R_t \in M_{N \times 1}, \quad f_t \in M_{k \times 1}, \quad b \in M_{N \times k}, \quad (9.80)$$

where R_t is a vector of N asset returns, f_t is a vector of k common factors, and b is an $N \times k$ matrix of factor loadings. If factor returns follow this formulation, then the arbitrage pricing equilibrium implies that the expected return on the vector of assets is a linear function of the factor loadings

$$E[R] = \lambda_0 + b\lambda'. \quad (9.81)$$

The question then becomes: How to identify the common factors? One approach is to let the stock returns determine the factors. Given the linear factor model above, the variance matrix for the returns on the vector of assets becomes

$$\Sigma = \beta\beta' + \theta, \tag{9.82}$$

where θ is a diagonal matrix. Under this specification, the vector of factor loadings can be estimated by maximizing

$$L = -\ln|\beta\beta' + \theta| - \text{tr}[S(\beta\beta' + \theta)^{-1}], \tag{9.83}$$

where $\text{tr}[\cdot]$ denotes the trace of the matrix. Based on this estimated vector of factor loadings, the risk premium for each coefficient are presented in Table 9.6.

The results of the basic factor model presented in Eq. (9.81) are presented in Table 9.7. Augmenting the model to test for disequilibria

Table 9.6. Estimated factor loadings.

Stock number	Factor 1	Factor 2
11499	0.00576	−0.05389
12140	0.03512	−0.06152
15553	0.02569	−0.00286
18542	0.01408	−0.05615
19166	0.01922	−0.05780
19749	−0.01091	−0.00215

Table 9.7. Estimated risk premia.

Parameter	Estimate
λ_0	0.01123
	(0.00200)
λ_1	0.00397
	(0.05674)
λ_2	−0.07589
	(0.03665)

Table 9.8. Test for agribusiness risk premia.

Parameter	Estimate
λ_0	0.01061
	(0.00198)
λ_1	−0.00940
	(0.05568)
λ_2	−0.08497
	(0.03601)
λ_3	0.00413
	(0.00229)

in the equity market for agribusinesses

$$E[R_i] = \lambda_0 + \lambda_1 b_{1i} + \lambda_2 b_{2i} + \lambda_3 D_i. \tag{9.84}$$

The estimates of the augmented APT formulation are presented in Table 9.8. Again, the APM results indicate that the returns for agribusiness stocks do not possess a nonsystematic component. Put slightly differently, agribusiness stocks are in arbitrage equilibrium with the remainder of the stock market.

9.6. Summary

Following economic insights, differences in risk should imply differences in market prices for investments. This chapter presents two related market models for risk: the CAPM and the APM. Empirically, these models can be used to estimate whether the capital market is efficient (which is to say that the asset market reflects the relative risk for each investment) and to adjust traditional investment models.

Chapter 10

Option Pricing Approaches to Risk

An option is of one a variety of assets known as derivatives, that is, an asset that derives its value from another asset. One example is the *stock option* where the option gives the holder the right to purchase a stock of some company at a stated price. Historically, options have entered agricultural economics as a market tool (i.e., an alternative to hedging as discussed below). However, in the early 1990s, options literature expanded beyond financial and marketing instruments with the emergence of *real options* literature. In real options literature, researchers use the option pricing approach to value assets such as investment opportunities. This chapter begins by developing the options from a traditional market perspective and then extends the development to include the valuation of real options.

10.1. Introduction to Options and Futures

In order to develop the options instrument, this section starts with a discussion of the futures market. Traditionally, researchers and extension professionals in the United States suggested that hedging using futures instruments could be used to reduce output price risks in the farm sector.

10.1.1. *Futures and the Hedge*

Most undergraduate courses on agricultural marketing include a section on futures markets and hedging. However, to discuss options and option pricing, this chapter briefly introduces the use of futures markets. Futures markets such as the Chicago Board of Trade allow for the trading (purchase and sale) of commodities to be delivered at some future data. On November 18, 2004, the future price for December 2004 corn was 204'0 ($2.04 and 0/8).

From a risk management standpoint, decision makers could choose to forward market cattle in November on August 2, 2004. Forward marketing using futures markets is typically referred to as hedging. To forward market cattle on August 2, 2004 for sale on November 18, 2004, the farmer would sell a contract of fat cattle on August 2. Referring to Table 10.1, the price of fat cattle in the August 2 for January–February delivery is $89.62/cwt.

When the farmer marketed cattle on November 18, 2004, he received $85.44/cwt on the cash market. In addition, he gained $0.72/cwt on the futures market (the price that he buys back the futures contract for was $88.90/cwt). If the farmer had attempted to market his cattle on June 17, 2004, he would have lost $0.30/cwt on each contract resulting in a $85.14/cwt price.

In general, the return on a hedge is based on the expected *basis*, which is defined as the difference between the *cash price* and the futures price on the date that the futures transaction is reversed.

$$B_t = F_t - C_t \qquad (10.1)$$

Table 10.1. Futures prices for fat cattle.

Date	Open	High	Low	Close
11/17/2004	88.05	89.10	87.95	88.90
8/2/2004	89.30	90.02	89.25	89.62
6/17/2004	88.45	88.75	88.15	88.60
Transactions				
Cash price	85.44	Sell on 8/2		89.62
		Buy on 11/17		88.90
		Gain		0.72
		Net price		86.16

where B_t is defined as the basis, F_t is the price in the futures market, and C_t is the price in the cash market. Rearranging this expression

$$C_t = F_t - B_t \qquad (10.2)$$

or taking the expectation

$$E(C_t) = E(F_t) - E(B_t). \qquad (10.3)$$

So the expected cash price at the date the hedge is initiated (on August 2, 2004) less the expected basis [$E(B_t)$]. Any risk in the price then comes from the risk of the basis.

10.1.2. *Options*

What is an option? In a general sense, an option is exactly what its name implies — an option is the opportunity to buy or sell one share of stock or lot of commodity at some point in the future at some stated price. For example, a *call* option entitles the purchaser the right to purchase a stock or commodity at some stated price. From Table 10.2, the right to purchase fat cattle at $88.00/cwt in December costs $2.60/cwt. On the flip side, the instrument that gives the bearer the right to sell a stock or unit of commodity at a stated price is called a *put*.

Options are *contingent assets*, they only have a value contingent on certain outcomes in the economy. For example, if a producer purchases a call with a strike price of $88.00/cwt for $2.60 and the cattle price increases to $92.00/cwt next month, he would exercise the option grossing $4.00/cwt ($92.00/cwt–$88.00/cwt). The net return is $1.40/cwt ($4.00/cwt–$2.60/cwt). If, on the other hand, the price

Table 10.2. Price of fat cattle calls for December.

Strike	Open	High	Low	Last	Settle	Change
86	3.400	3.400	3.400	3.400	3.525	−1650
88	2.500	2.750	2.500	2.700	2.600	−1300
89	2.400	2.400	2.400	2.400	2.200	−1200
90	1.600	1.900	1.600	1.600	1.700	−1100

of cattle decreases to \$84.00/cwt in December, the option becomes valueless. Exercising the option would result in a \$4.00/cwt gross loss (\$84.00/cwt–\$88.00/cwt) and a total loss of \$6.60 (\$4.00/cwt from exercising the option and \$2.60/cwt on the purchase price of the option).

Graphically, the payoff function for an option is depicted in Fig. 10.1. Technically, there are two types of options: a *European option* and an *American option*. A European option can be only exercised on the expiration date. The American option can be exercised on any date up until the expiration date.

Given these differences, let $F(s, t; T, x)$ denote the value of an American call option with stock price s on date t with an expiration date of T for an exercise price of x. The European call option is denoted $f(s, t; T, x)$. Where $G(s, t; T, x)$ is an American put option and $g(s, t; T, x)$ is the European put option (this notation largely follows the material developed by Ingersoll, 1987). Based on these definitions, the American option is generally more difficult to value because s could become greater than x at any time before T. In addition, the investor may not wish to exercise the option at the instant that s exceeds x because there might be more money to be made if the stock price continues to rise.

Risk neutral propositions simply assume that investors prefer more to less.

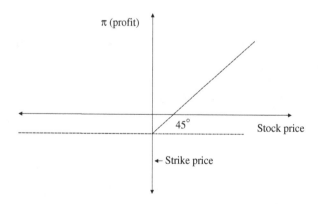

Fig. 10.1.   Payoff of a call option.

Proposition 10.1. $F(\cdot) \geq 0,\ G(\cdot) \geq 0, f(\cdot) \geq 0, g(\cdot) \geq 0.$
So that the value of American or European puts and calls are always nonnegative.

Proposition 10.2.

$$F(s, T; T, x) = f(s, T; T, x) = \max(s - x, 0)$$
$$G(s, T; T, x) = g(s, T; T, x) = \max(x - s, 0).$$

At the point of termination, the price of American puts or calls is equal to the value of European puts and calls that are equal to the difference in value between the strike (or exercise) price and the stock price

Proposition 10.3.

$$F(s, t; T, x) \geq s - x$$
$$G(s, t; T, x) \geq x - s.$$

The value of the American put or call must exceed the difference between the strike and stock prices at any time period before the exercise date.

Proposition 10.4.   For $T_2 > T_1$

$$F(.; T_2, x) \geq F(.; T_1, x)$$
$$G(.; T_2, x) \geq G(.; T_1, x).$$

The longer the time to the exercise date, the higher the price of an American put or call.

Proposition 10.5. $F(\cdot) \geq f(\cdot)$ and $G(\cdot) \geq g(\cdot).$
An American put or call is always more valuable than an European put or call.

Proposition 10.6.   For $x_1 > x_2$

$$F(.; x_1) \leq F(.; x_2) \quad \text{and} \quad f(.;, x_1) \leq f(.; x_2)$$
$$G(.; x_1) \geq G(.; x_2) \quad \text{and} \quad g(.; x_1) \geq g(.; x_2).$$

The higher the stock price, the lower the value of both American and European calls and the higher the value of both American and European puts.

Proposition 10.7.

$$s = F(s,t;\infty,0) \geq F(s,t;T,x) \geq f(s,t;T,x).$$

The first equality involves the definition of a stock in a limited liability economy. If you purchase a stock, you purchase the right to sell the stock between now and infinity. Further, given limited liability, you will not sell the stock for less than zero.

Proposition 10.8. $f(0,.) = F(0,.) = 0.$

Intuitive determinants of European option prices: Restating three of the propositions: (1) the value of a call option is an increasing function of the spot stock price (S); (2) the value of a call option is a decreasing price of the strike price (X); (3) the value of a call option is an increasing function of the *time to maturity* (T). In addition, the value of an option is an increasing function of the variability of the underlying asset (Fig. 10.2). To see this, think about imposing the probability density function over a "zero price" option.

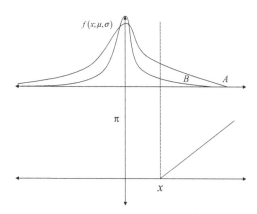

Fig. 10.2. The value of a call option based on changes in the variance of the underlying asset.

10.1.3. *Option Pricing using Black–Scholes*

The Black–Scholes model can be used to price European call options (Huang and Litzenberger, 1988). To develop this pricing mechanism, first construct the payoff function for a security on which the option is written as

$$x_j(k) = \begin{cases} x_j - k & \text{э: } x_j - k > 0 \\ 0 & \text{otherwise,} \end{cases} \tag{10.4}$$

where $x_j(k)$ is the payoff of a share of security j purchased at the exercise price of k. The price of the European call option is then defined as

$$p_j(S_j, k) \geq \max\left[S_j - \frac{k}{(1 + r_f)}, 0 \right]. \tag{10.5}$$

Consider the strategy of selling one share of the security to buy one European call option written on the security. The initial cost of the strategy is

$$p_j(S_j, k) - S_j + \frac{k}{(1 + r_f)}. \tag{10.6}$$

The possible payoffs of this strategy are

$$\begin{cases} x_j - k - x_j + k = 0 & \text{if } x_j \geq k \\ -x_j + k > 0 & \text{if } x_j < k. \end{cases} \tag{10.7}$$

In the first case, the decision maker would exercise the option buying back the stock at the original price while in the second case, the investor makes money because the stock price decreased (you make profit equal to the decrease in the stock price). Therefore, to avoid a risk-less profit (the investor cannot make something for nothing)

$$p_j(S_j, k) - S_j + \frac{k}{1 + r_f} > 0 \Rightarrow p_j(S_j, k) > S_j - \frac{k}{1 + r_f}. \tag{10.8}$$

Starting with a two-period economy, assume a power utility function for the representative agent

$$u_0(z_0) + u_1(z_1) = \frac{1}{1 - B} z_0^{1-B} + \rho \frac{1}{1 - B} z_1^{1-B}, \tag{10.9}$$

where ρ is the time preference parameter. The arbitrage condition (selling short a share of stock and purchasing a call option) then implies

$$p_j(S_j, k) = \rho E \left[\max[x_j - k, 0] \left(\frac{C}{C_0} \right)^{-B} \right]. \qquad (10.10)$$

Next, assume that x and C are lognormally distributed

$$\begin{pmatrix} \ln(x_j) \\ \ln(C) \end{pmatrix} \sim N \left(\begin{bmatrix} \mu_j \\ \mu_C \end{bmatrix}, \begin{bmatrix} \sigma_j^2 & \kappa \sigma_j \sigma_C \\ \kappa \sigma_j \sigma_C & \sigma_C^2 \end{bmatrix} \right), \qquad (10.11)$$

where κ is the correlation coefficient.

This assumption implies that $\ln(x_j / S_j)$ (the return on the short sale) and $\rho \ln(C/C_0)$ are normally distributed

$$\begin{pmatrix} \ln(\frac{x_j}{S_j}) \\ \rho \ln(\frac{C}{C_0}) \end{pmatrix}$$
$$\sim N \left(\begin{bmatrix} \mu_j - \ln(S_j) \\ -B\mu_C + \ln(\rho) + B\ln(C_0) \end{bmatrix}, \begin{bmatrix} \sigma_j^2 & -B\kappa\sigma_j\sigma_C \\ -B\kappa\sigma_j\sigma_C & B^2\sigma_C^2 \end{bmatrix} \right). $$
$$(10.12)$$

The value of the call option can then be written as

$$p_j(S_j, k) = S_j \int_{-\infty}^{\infty} \int_{\ln(\frac{k}{S_j})}^{\infty} \left(e^z - \frac{k}{S_j} \right) e^y f(z, y) dz \, dy, \qquad (10.13)$$

which can be rewritten as

$$p_j(S_j, k) = S_j \int_{-\infty}^{\infty} \int_{\ln(\frac{k}{S_j})}^{\infty} e^{z+y} f(z, y) dz \, dy$$
$$- k \int_{-\infty}^{\infty} \int_{\ln(\frac{k}{S_j})}^{\infty} e^y f(z, y) dz \, dy. \qquad (10.14)$$

Some mathematical niceties: dealing with the lower bound of the integral

$$\int_a^\infty f(z)dz \Rightarrow v = \frac{a-\mu}{\sigma} \Rightarrow \int_a^\infty f(z)dz = \int_{\frac{a-\mu}{\sigma}}^\infty n(v)dv, \quad (10.15)$$

which by the symmetry of the normal distribution function implies

$$\int_{\frac{a-\mu}{\sigma}}^\infty n(v)dv = \int_{-\infty}^{\frac{a-\mu}{\sigma}} n(v)dv = N\left(\frac{-1+\mu}{\sigma}\right). \quad (10.16)$$

Next, because of the geometric nature of the distribution function

$$\int_a^\infty e^z f(z)dz = \int_a^\infty \frac{1}{\sigma\sqrt{2\pi}}\exp\left[-\frac{1}{2\sigma^2}(z-\mu)^2 + z\right]dz$$

$$= (e^{\mu+\frac{\sigma^2}{2}})\int_a^\infty \frac{1}{\sigma\sqrt{2\pi}}\exp\left\{-\frac{1}{2\sigma^2}(z-[\mu-\sigma^2])^2\right\}dz$$

$$= (e^{\mu+\frac{\sigma^2}{2}})N\left(\frac{-a+\mu}{\sigma}+\sigma\right). \quad (10.17)$$

The conditional distribution

$$\int_{-\infty}^\infty e^y f(y\mid z)dy = \exp\left[\mu_c + \kappa\frac{\sigma_c}{\sigma_j}(z-\mu_j) + \frac{1}{2}(1-\kappa^2)\sigma_c^2\right]. \quad (10.18)$$

Combining these derivations

$$\int_{-\infty}^\infty \int_a^\infty e^y f(z,y)dz\,dy = \int_a^\infty f(z)\int_{-\infty}^\infty e^y f(y\mid z)dy\,dz$$

$$= \int_a^\infty e^{\mu_C + \kappa\frac{\sigma_C}{\sigma_j}(z-\mu_j) + (1-\kappa^2)\frac{\sigma_C^2}{2}} f(z)dz$$

$$= (e^{\mu_C + \frac{\sigma_C^2}{2}})\int_a^\infty \frac{1}{\sigma_j\sqrt{2\pi}}$$

$$\times \exp\left[-\frac{1}{2\sigma_j^2}(z-(\mu_j-\kappa\sigma_j\sigma_C))^2\right]dz. \quad (10.19)$$

Now back to the original integral

$$p_j(S_j, k) = S_j \int_{-\infty}^{\infty} \int_{\ln(\frac{k}{S_j})}^{\infty} e^{z+y} f(z, y) dz\, dy$$

$$- k \int_{-\infty}^{\infty} \int_{\ln(\frac{k}{S_j})}^{\infty} e^y f(z, y)\, dz\, dy. \qquad (10.20)$$

Taking the last part first

$$\int_{-\infty}^{\infty} \int_a^{\infty} e^y f(z, y) dz\, dy = \left(e^{\mu_C + \frac{\sigma_C^2}{2}}\right) N\left(\frac{-a + \mu_j}{\sigma_j} + \kappa\sigma_C\right). \qquad (10.21)$$

Taking the first part

$$\int_{-\infty}^{\infty} \int_a^{\infty} e^{z+y} f(z, y) dz\, dy = \left(e^{\mu_j + \mu_C + \frac{(\sigma_j^2 + 2\kappa\sigma_j\sigma_C + \sigma_C^2)}{2}}\right)$$

$$\times N\left(\frac{-a + \mu_j}{\sigma_j} + \kappa\sigma_C + \sigma_j\right). \qquad (10.22)$$

To finish the derivation, we assume

$$e^{\mu_C + \frac{1}{2}\sigma_C^2} = \frac{1}{1 + r_f} \qquad (10.23)$$

or that future consumption is discounted at the risk-free rate of return. In addition, assume

$$e^{\mu_j + \mu_C + \frac{\left(\sigma_j^2 + 2\kappa\sigma_j\sigma_C + \sigma_C^2\right)}{2}} = 1, \qquad (10.24)$$

which is implicitly the pricing condition of stock in period 0 given its utility distribution in period 1 (enforces a zero arbitrage condition on the stock price). And defining

$$Z \equiv \frac{\ln(\frac{S_j}{k}) + (\mu_j + \kappa\sigma_j\sigma_C)}{\sigma_j}. \qquad (10.25)$$

The option price becomes

$$p_j(S_j, k) = S_j N(Z + \sigma_j) - \frac{1}{1 + r_f} k N(Z). \tag{10.26}$$

Following an Ito's Lemma formulation, the equation of motion for the stochastic process can be expressed as

$$\Delta E = \mu \Delta t + \sigma \Delta z. \tag{10.27}$$

Defining the Wiener increment [following Kamien and Schwartz's (1991) definitions]

$$\Delta z = \varepsilon_t \sqrt{\Delta t}. \tag{10.28}$$

The expectation and the variance of the equation of motion for equity can then be defined as

$$\begin{aligned} \text{Exp}[\Delta E] &= \mu \Delta t \\ V[\Delta E] &= \sigma^2 \Delta t. \end{aligned} \tag{10.29}$$

Thus, the Black and Scholes result can be rewritten using stochastic process as

$$c = SN(d_1) - Xe^{-r_f T} N(d_2)$$
$$d_1 = \frac{\ln(\frac{S}{X}) + r_f T}{\sigma \sqrt{T}} + \frac{1}{2} \sigma \sqrt{T} \tag{10.30}$$
$$d_2 = d_1 - \sigma \sqrt{T}.$$

Table 10.3 presents the price of a call option for different strike prices computed using the Black–Scholes pricing formula. Consistent with expectations, the price of the call option is declining in the strike price.

10.2.   Real Option Valuation

Traditional courses in financial management state that an investment should be undertaken if the net present value of the investment is positive. However, firms routinely fail to make investments that appear

Table 10.3. Example of Black–Scholes pricing.

Strike price	Price today	Weeks to maturity	Std. Dev.	T	d_1	d_2	Price
87.00	89.00	1.0	0.150	0.019	1.126	1.105	2.164
88.00	89.00	1.0	0.150	0.019	0.577	0.556	1.370
89.00	89.00	1.0	0.150	0.019	0.034	0.013	0.760
90.00	89.00	1.0	0.150	0.019	−0.504	−0.524	0.360
91.00	89.00	1.0	0.150	0.019	−1.035	−1.056	0.143
92.00	89.00	1.0	0.150	0.019	−1.560	−1.581	0.047
87.00	89.00	10.0	0.150	0.192	0.452	0.386	3.711
88.00	89.00	10.0	0.150	0.192	0.278	0.212	3.097
89.00	89.00	10.0	0.150	0.192	0.106	0.040	2.549
90.00	89.00	10.0	0.150	0.192	−0.064	−0.130	2.069
91.00	89.00	10.0	0.150	0.192	−0.232	−0.298	1.656
92.00	89.00	10.0	0.150	0.192	−0.398	−0.464	1.306

profitable considering the time value of money. Several alternative explanations for this phenomenon have been proposed. However, the most fruitful involves risk.

Integrating risk into the decision model may take several forms from the capital asset pricing model to stochastic net present value. However, one avenue, which has gained increased attention during the past decade, is the notion of an investment as an option.

Several characteristics of investments make the use of option pricing models attractive. In most investments, investors can be construed to have limited liability with the distribution being truncated at the loss of the entire investment. Alternatively, Dixit and Pindyck have pointed out that the investment decision is very seldom a now or never decision. The decision maker may simply postpone exercising the option to invest.

10.2.1. *Derivation of the Value of Waiting*

As a first step in the derivation of the value of waiting, consider an asset whose value changes over time according to a geometric Brownian motion stochastic process

$$dV = \alpha V \, dt + \sigma V \, dt. \tag{10.31}$$

Given the stochastic process depicting the evolution of asset values over time, assume that there exists a perfectly correlated asset that obeys a similar process

$$dx = \mu x \, dt + \sigma x \, dz$$

$$\mu = r + \phi \rho_{vm} \sigma. \tag{10.32}$$

Comparing the two stochastic processes leads to a comparison of α and μ. The relationship between these two values gives rise to the execution of the option. Defining $\delta = \mu - \alpha$ to the dividend associated with owning the asset. α is the capital gain while μ is the "operating" return. If δ is less than or equal to zero, the option will never be exercised. Thus, $\delta > 0$ implies that the operating return is greater than the capital gain on a similar asset.

Next, construct a risk-free portfolio containing one unit of the option to some level of short sale of the original asset

$$P = F(V) - F_V(V)V \tag{10.33}$$

P is the value of the risk-free portfolio, $F(V)$ is the value of the option, and $F_V(V)$ is the derivative of the option price with respect to value of the original asset. Dropping the Vs and differentiating the risk-free portfolio determine the rate of return on the portfolio. To this differentiation, append two assumptions, (1) the rate of return on the short sale over time must be $-\delta V$ (the short sale must pay at least the expected dividend on holding the asset), and (2) the rate of return on the risk-free portfolio must be equal to the risk-free return on capital $r(F - F_V V)$

$$dF - F_V \, dV - \delta V F_V \, dt = r(F - F_V V) \, dt. \tag{10.34}$$

Combining this expression with the original geometric process and applying Ito's Lemma, derive the combined zero-profit and zero-risk condition

$$\frac{1}{2}\sigma^2 V^2 F_{VV} + (r - \delta)VF_V - rF = 0. \qquad (10.35)$$

In addition to this differential equation, there are three boundary conditions

$$F(0) = 0, \qquad F(V^*) = V^* - I, \qquad F_V(V^*) = 1. \qquad (10.36)$$

The solution of the differential equation with the stated boundary conditions is

$$F(V) = \alpha V^\beta \qquad (10.37)$$

$$\alpha = \frac{(V^* - I)}{V^{*\beta}} \qquad (10.38)$$

$$V^* = \frac{\beta}{(\beta - 1)}I \qquad (10.39)$$

$$\beta = \frac{1}{2} - \frac{(r - \delta)}{\sigma^2} + \left\{\left[\frac{(r - \delta)}{\sigma^2} - \frac{1}{2}\right]^2 + 2\frac{r}{\sigma^2}\right\}^{\frac{1}{2}} \qquad (10.40)$$

β then simplifies to

$$\beta = \frac{1}{2}\left\{1 + \left(1 + \frac{8r}{\sigma^2}\right)^{\frac{1}{2}}\right\}. \qquad (10.41)$$

Estimating β, in order to incorporate risk into an investment decision using the Dixit and Pindyck approach, σ must be estimated.

This approach to estimating σ is through simulation. Specifically, simulating the stochastic Net Present Value of an investment as

$$V_t = \sum_{i=1}^{N+t} \frac{CF_i}{(1 + r)^i}. \qquad (10.42)$$

Converting this value to an infinite streamed investment then involves

$$APV_t = \frac{V_t}{\left(\frac{1-\frac{1}{(1+r)^{-N}}}{r}\right)} \tag{10.43}$$

$$V_t^* = \frac{APV_t}{r}. \tag{10.44}$$

The parameters of the stochastic process can then be estimated by

$$\frac{dV}{V} = d\ln(V)$$

$$= \ln(V_{t+1}) - \ln(V_t). \tag{10.45}$$

10.2.2. *Application to Citrus*

The simulated results indicate that the present value of orange production was \$852.99/acre with a standard deviation of \$179.88/acre (Moss, Pagano, and Boggess, 1994). Clearly, this investment is not profitable given an initial investment of \$3,950/acre. The average log change based on 7500 draws was 0.0084693 with a standard deviation of 0.0099294.

Assuming a mean of the log change of zero, the computed value of β is 25.17 implying a $\beta/(\beta - 1)$ of 1.0414. Hence, the risk adjustment raises the hurdle rate to \$4113.40. Alternatively, the value of the option to invest given the current scenario is \$163.40.

10.3. Crop Insurance

The general concept of insurance is the construction of an instrument or gamble that pays the purchaser in the event of some adverse occurrence. Frequently, purchased insurance contracts include life insurance, which pays in the event of the holders death, car insurance, which pays in the case of an accident, catastrophic health insurance, which pays in the event of a major medical event such as cancer, etc. Each of these contracts specifies a payable event, an indemnity (the amount to be paid on the event) and a premium (the amount paid for the insurance contract).

Under commercial insurance arrangements, the premium charged for the insurance is generally considered to be actuarially sound. Specifically, the expected indemnity payments are exactly equal to premiums charged. If the premiums exceeded the expected indemnity payments, then insurance firms would earn abnormal profits. These abnormal profits would be bid out of the market by new firms entering the insurance arena. If premiums fell short of the expected indemnity, the insurance firm would lose money and ultimately exit the industry.

The actuarial value of an insurance contract can then be written as

$$V = \int_0^{y^*} Pyf(y)dy, \qquad (10.46)$$

where V is the value of crop yield insurance, P is the price of the crop, y is the variable of integration, $f(y)$ is the probability density function for crop yields, and y^* is the minimum insured yield (trigger yield in crop insurance).

Current debates in the area of crop yield insurance involve estimation of the probability density function for yields, $f(y)$. Most common statistical applications assume that the probability density function is normal or asymptotically normal. This assumption may have serious shortcomings in the valuation of crop insurance. From an agronomic perspective, yields are bounded by zero on the downside and limiting nutrients such as nitrogen on the up side. Hence, at the least, the $x \in (-\infty, \infty)$ of the normal distribution would appear to be violated. However, the truncated normal distribution may be appropriate for crop yields. The debate of potential normality of crop yields typically revolves around skewness and kurtosis. Skewness is a measure of non-symmetry of the distribution. The normal distribution is symmetric and, hence, yields zero skewness. A significant portion of the literature supports skewness in yields, but as pointed out by Just and Weninger (1999), it does not reach a consensus on the direction of skewness. Kurtosis measures the relationship between the area in the tails and the area around the means.

The second area of debate in the area of crop insurance is the moral hazard/incentive compatibility dimension of crop insurance. A basic problem in any insurance contract is the determination of the insurable event and the amount of damages. The classic scenario of filing lawsuits for pain and suffering after a car accident due to whiplash is a famous lawyer joke. In the current case, the insurance company must determine when a yield reduction has occurred and the amount of that reduction. A second problem is the difficulty of self-selection. Specifically, as in health insurance contracts, riskier farmers will be willing to pay more money for insurance than safer farmers.

Using the data from Ramirez, Moss, and Boggess, derive the parameters of the normal distribution function for corn as $\mu = 173.03$, $\sigma = 8.71$

$$V = \int_0^{y^*} py \frac{1}{8.71\sqrt{2\pi}} \exp\left[-\frac{(y - 173.03)^2}{151.88} \right] dy. \tag{10.47}$$

Assuming a corn price of $3.00/bushel, the value of insurance becomes the insurance premiums presented in Table 10.4.

Integrating price insurance, in order to derive the price of risk the actuarial premium becomes

$$V = \int_0^{p^*} \int_0^{y^*} pyf(y, p) dy\, dp. \tag{10.48}$$

The joint distribution is specified using the futures price as an efficient estimate of the price at harvest time. The price at harvest time can be estimated as a function of the futures price at planting

$$p_t^b = \alpha_0 + \alpha_1 f_{t-1}^b + \varepsilon_t. \tag{10.49}$$

Table 10.4. Insurance premiums.

Coverage	Insurance premiums
0.90×173.03	10.73
0.85×173.03	0.63
0.80×173.03	0.01

Given traditional assumptions, α_0 is the anticipated basis and α_1 is equal to one. The distribution of price is then a function of the distribution of ε_t.

10.4.   Summary

This chapter presents a variety of models of decision making under risk typically referred to as derivatives or options. In each case, the value of the instrument is dependent on the value of another asset. In the financial markets, the price of a put or a call is dependent on the stock price. In the real options literature, the value of an investment opportunity today is at least in part a function of what the value of the investment opportunity may be tomorrow. Finally, the value of a crop insurance instrument is contingent on the crop yield.

Chapter 11

State Contingent Production Model: The Stochastic Production Set

So far this book has focused on the risk formulations where the input decisions were taken as given (i.e., not considered in the analysis of risk). Put differently, the overall scale and scope of a variety of investment alternatives are identified and then the decision maker chooses between the fixed alternatives. Chapter 3 derived the certainty equivalent and risk premium for individual investment opportunities. Chapter 5 presented the consequences of holding different portfolios of investment opportunities while the returns and risks of those opportunities are taken as a given. Chapter 6 presented a variety of farm-planning models where the decision maker chooses among different crop opportunities to control the aggregate risk facing the firm. Similarly, Chapter 7 demonstrated how risk efficiency criteria can be used to select among different alternatives with different risk profiles. This chapter examines the interaction between production input decisions and risk using the state-contingency model proposed by Chambers and Quiggin (2000).

11.1. Depicting Risk and Input Decisions in the Production Function

Historically, most applications of risk in production theory have been based on a *stochastic production function*. In its simplest format, the stochastic production function is simply a production function with a random error term added to it:

$$y = f(x) = x_1^\alpha x_2^\beta + \varepsilon \tag{11.1}$$

where y is the output level, $f(x)$ is the production function, x_1 and x_2 are inputs, and ε is a random error representing risk. Under this formulation, the risk (dispersion) created by the error term did not affect the input decisions. In a slightly more complex scenario, the simple formulation can be expanded to allow the risk to be a function of one or both inputs:

$$y = \tilde{f}(x) = x_1^\alpha x_2^\beta + \varepsilon(x_1, x_2), \tag{11.2}$$

where $\tilde{f}(x)$ is the stochastic production function since the risk is now a function of the amount of each input applied. This production function follows the formulation proposed by Just and Pope (1978). The specification allows inputs to increase or decrease the risk of production. However, even with the expansion of the error function to include relative effects of each input, the assumptions used to parametrize the model imply that the ratio of input use did not change.

Adapting this formulation slightly, Chambers and Quiggin depict this original production function as dependent on a set of deterministic inputs and a stochastic input:

$$z = f(x, \varepsilon), \tag{11.3}$$

where z is the level of output, which is a function of *controllable inputs* (x) and *uncontrollable factors* (ε). Simplifying this representation, we assume that the random variable is a Bernoulli variable that can take values $\varepsilon = \{0, 1\}$. The production function for this representation is depicted in Fig. 11.1. The random variable has an effect that is largely independent of input choice or the introduction of the random variable simply shifts the frontier upward.

11.1.1.   *Estimation of Stochastic Production Functions using Quantile Regression*

One formulation for the stochastic production function that is consistent with the state contingency model can be derived from a Quantile

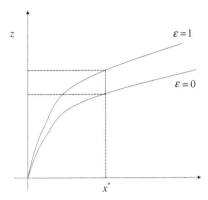

Fig. 11.1.   Stochastic production function.

Regression as proposed by Koenker and Bassett (1978). As a starting point for this discussion, recall the general mapping function formulation used in Chapter 2 in the development of random variables.

In the stochastic production function specification, the production function is written as a combination of random and non-random or choice variables as

$$f : X \times \Omega \rightarrow R_+^1, \tag{11.4}$$

where X denotes the set of controlled variables, Ω is the set of all possible outcomes of the uncontrollable or random factors of production, and R_+^1 depicts the set of all positive real numbers. Assuming that this mapping obeys the rules of a function (i.e., a unique output is associated with each pair of controllable and uncontrollable variables), the function can be approximated using a second-order Taylor series expansion around a fixed point $(x_0, \omega_0) \in (X, \Omega)$

$$f(x, \omega) = f(x_0, \omega_0) + \nabla_{(x\omega)} f(x_0, \omega_0) \left(\begin{bmatrix} x \\ \omega \end{bmatrix} - \begin{bmatrix} x_0 \\ \omega_0 \end{bmatrix} \right)$$

$$+ \frac{1}{2} \left(\begin{bmatrix} x \\ \omega \end{bmatrix} - \begin{bmatrix} x_0 \\ \omega_0 \end{bmatrix} \right)' \nabla^2_{(x\omega)} f(x_0, \omega_0) \left(\begin{bmatrix} x \\ \omega \end{bmatrix} - \begin{bmatrix} x_0 \\ \omega_0 \end{bmatrix} \right)$$

$$+ O^2(x_0, \omega_0), \tag{11.5}$$

where the last term is the error in the approximation, which can be taken to be a small constant converging to zero.

Solving the gradient and Hessian matrix at the point of approximation yields

$$f(x_0, \omega_0) = \alpha_0$$
$$\nabla_{(x\omega)} f(x_0, \omega_0) = \alpha(x_0, \omega_0) = (\alpha_x(x_0, \omega_0) \alpha_\omega(x_0, \omega_0))$$
$$\nabla^2_{(x\omega)} f(x_0, \omega_0) = A(x_0, \omega_0) = \begin{pmatrix} A_{xx}(x_0, \omega_0) & A_{x\omega}(x_0, \omega_0) \\ A_{\omega x}(x_0, \omega_0) & A_{\omega\omega}(x_0, \omega_0) \end{pmatrix}, \tag{11.6}$$

where the subscripts denote a conformable partition of each vector or matrix. Substituting Eq. (11.6) into Eq. (11.5) and carrying out the matrix operations yields

$$\begin{aligned} f(x, \omega) &= \alpha_0 + \alpha_x(x_0, \omega_0)(x - x_0) + \alpha_\omega(x_0, \omega_0)(\omega - \omega_0) \\ &+ \frac{1}{2}(x - x_0)' A_{xx}(x_0, \omega_0)(x - x_0) \\ &+ (x - x_0)' A_{\omega x}(x_0, \omega_0)(\omega - \omega_0) \\ &+ \frac{1}{2}(\omega - \omega_0)' A_{\omega\omega}(x_0, \omega_0)(\omega - \omega_0) \\ &+ O^2(x_0, \omega_0). \end{aligned} \tag{11.7}$$

Expressing this function as a deterministic function plus an error term

$$f(x, \omega) = g(x, \omega) + h(x, \omega)$$
$$g(x, \omega) = \alpha_0 + \alpha_x(x_0, \omega_0)(x - x_0) + \frac{1}{2}(x - x_0)' A_{xx}(x_0, \omega_0)(x - x_0)$$
$$\begin{aligned} h(x, \omega) &= \alpha_\omega(x_0, \omega_0)(\omega - \omega_0) + (x - x_0)' A_{\omega x}(x_0, \omega_0)(\omega - \omega_0) \\ &+ \frac{1}{2}(\omega - \omega_0)' A_{\omega\omega}(x_0, \omega_0)(\omega - \omega_0) + O^2(x_0, \omega_0) \end{aligned} \tag{11.8}$$
$$\begin{aligned} &= \Big[\alpha_\omega(x_0, \omega_0) + (x - x_0)' A_{\omega x}(x_0, \omega_0) \\ &+ \frac{1}{2}(\omega - \omega_0)' A_{\omega\omega}(x_0, \omega_0) \Big] (\omega - \omega_0) \\ &+ O^2(x_0, \omega_0), \end{aligned}$$

where $g(x, \omega)$ is the *deterministic component* of the production function and $h(x, \omega)$ is the random term of the Just and Pope production function.

This formulation allows for two different types of randomness in the production function. If $A_{\omega x}(x_0, \omega_0) = 0$, there is no interaction between the input use and the variability of production (i.e., inputs are neither risk increasing nor risk reducing). If $A_{\omega x}(x_0, \omega_0) < 0$, the input is risk reducing, while if $A_{\omega x}(x_0, \omega_0) > 0$, the input is risk increasing.

This raises some interesting questions about the specification of the error terms. For example, consider two possible specifications of the error terms: (1) the assumption that errors are normally distributed or (2) the assumption that the error is distributed negative exponential. In both cases, we will assume that $\omega_0 \to 0$. Under the assumption of normality, the typical assumption is that $\omega \sim N(0, \sigma_\omega^2)$. Hence taking the expansion of Eq. (11.8) around the mean yields a symmetric variation around the production function. However, the assumption that the random error has a negative exponential distribution yields a specification consistent with the technical frontier found in technical efficiency formulations. In the negative exponential specification, $\omega \to 0$ yields the efficient frontier. All other values yield outcomes of ω below the frontier.

To simplify the general stochastic frontier model in Eq. (11.8), consider a production function with two controllable inputs and one random outcome (i.e., a stochastic production function for wheat where x_1 is the quantity of nitrogen applied per acre, x_2 is the quantity of phosphorous applied per acre, ω is a single random factor, which is only observed after it occurs such as rainfall, and y is the level of hard red winter wheat produced). If ω could be observed, the stochastic production function could be specified as

$$\tilde{f}(x_1, x_2, \omega) = \alpha_0 + \begin{pmatrix} \alpha_1 \\ \alpha_2 \\ \alpha_3 \end{pmatrix}' \begin{pmatrix} x_1 \\ x_2 \\ \omega \end{pmatrix}$$

$$+ \frac{1}{2} \begin{pmatrix} x_1 \\ x_2 \\ \omega \end{pmatrix}' \begin{pmatrix} A_{11} & A_{12} & A_{13} \\ A_{12} & A_{22} & A_{23} \\ A_{13} & A_{23} & A_{33} \end{pmatrix} \begin{pmatrix} x_1 \\ x_2 \\ \omega \end{pmatrix}. \quad (11.9)$$

Equation (11.9) can be rewritten as

$$\tilde{f}(x_1, x_2, \omega) = \alpha_0 + \begin{pmatrix} \alpha_1 \\ \alpha_2 \end{pmatrix}' \begin{pmatrix} x_1 \\ x_2 \end{pmatrix} + \frac{1}{2} \begin{pmatrix} x_1 \\ x_2 \end{pmatrix}' \begin{pmatrix} A_{11} & A_{12} \\ A_{12} & A_{22} \end{pmatrix} \begin{pmatrix} x_1 \\ x_2 \end{pmatrix}$$

$$+ \omega \left(x_1 A_{13} + x_2 A_{23} + \frac{1}{2} \omega A_{33} \right)$$

$$= \alpha_0 + \begin{pmatrix} \alpha_1 + \omega A_{13} \\ \alpha_2 + \omega A_{23} \end{pmatrix}' \begin{pmatrix} x_1 \\ x_2 \end{pmatrix}$$

$$+ \frac{1}{2} \begin{pmatrix} x_1 \\ x_2 \end{pmatrix}' \begin{pmatrix} A_{11} & A_{12} \\ A_{12} & A_{22} \end{pmatrix} \begin{pmatrix} x_1 \\ x_2 \end{pmatrix} + \omega^2 A_{33}. \quad (11.10)$$

Following the discussion of Eq. (11.8), if $A_{13} = A_{23} = 0$, the choice of input level does not affect the overall risk (so that the inputs are neither risk increasing nor risk reducing). However, the quadratic approximation places significant restrictions on the specification. Specifically, the interaction between risk and inputs enters linearly (i.e., the intercept of the value of marginal product curves shift with the uncontrollable variable).

A more flexible formulation is to estimate the quadratic production function using a *quantile regression* formulation. In this formulation, the linear and quadratic terms differ depending on the quantile of the population estimated

$$P(Y_i < y) = F(y - x_i \beta), \quad (11.11)$$

where $P(Y_i < y)$ is the probability that the crop yield (wheat) is below a specified level and $F(y - x_i \beta)$ is a cumulative probability density function, which is conditioned on the estimated parameters β of the quadratic production function. Koenker and Bassett demonstrate that this regression relationship for the θth quantile can be estimated by solving

$$\min_{\beta \in R^k} \left[\sum_{i \in \{i: y_i \geq x_i \beta\}} \theta | y_i - x_i \beta | + \sum_{i \in \{i: y_i < x_i \beta\}} (1 - \theta) | y_i - x_i \beta | \right], \quad (11.12)$$

where there are k independent variables including the constant. In this specification, both the constant terms (αs in Eq. (11.10)) and the quadratic terms (the *A*s in Eq. (11.10)) depend on the random outcome.

Table 11.1 presents the estimated production function for the 0.20, 0.50, and 0.80 quantiles of the quadratic production function for hard

Table 11.1. Quantile regression results for wheat stochastic production function.

Variable	Coefficients	Lower bound	Upper bound
0.20 Quantile			
(Intercept)	19.713	−14.546	34.083
Nitrogen	0.278	0.080	0.666
Phosphorous	−0.544	−1.110	0.600
Nitrogen^2	−0.006	−0.008	−0.004
Phosphorous^2	−0.005	−0.017	−0.004
N × P	0.015	0.006	0.041
Missouri	7.727	2.103	9.246
Nebraska	10.241	8.166	13.062
Oklahoma	13.504	9.835	14.416
0.50 Quantile (Median)			
(Intercept)	9.647	−5.056	24.792
Nitrogen	0.325	−0.038	0.616
Phosphorous	0.097	−1.428	1.052
Nitrogen^2	−0.006	−0.008	−0.002
Phosphorous^2	−0.013	−0.023	0.020
N × P	0.015	−0.002	0.023
Missouri	8.028	6.096	9.743
Nebraska	10.665	8.456	12.439
Oklahoma	11.486	10.666	13.202
0.80 Quantile			
(Intercept)	32.906	13.980	60.968
Nitrogen	0.484	−0.216	0.844
Phosphorous	−1.217	−2.762	0.822
Nitrogen^2	−0.004	−0.008	0.000
Phosphorous^2	0.009	−0.034	0.040
N × P	0.007	−0.003	0.031
Missouri	9.106	5.296	11.372
Nebraska	11.181	8.025	16.968
Oklahoma	13.233	8.806	14.811

N-Nitrogen
P-Phosphorous

red winter wheat in Kansas, Missouri, Nebraska, Oklahoma, and Texas based on nitrogen and phosphorous. Following the preceding discussion, there are changes in both the linear terms and the quadratic terms that can be used to derive state dependent production models. This state dependency implies that the optimal amount of each fertilizer varies depending on the state of the uncontrollable variable.

11.1.2. *Developing the State-Space Representation*

The stochastic production function specification in Eqs. (11.2) and (11.3) is similar to the discussion of disembodied technological change. A disembodied technological change is a change that does not introduce a factor bias or input effect. Instead of this approach, Chambers and Quiggin map (or depict) the interaction between input use and random event in a different space (*state-contingent output space*). In this space, a single production choice implies two different state-spaces depicted in Fig. 11.2. Intuitively, different production choices yield different points as depicted in Fig. 11.3.

These choices are called acts (a) that have different consequences (y) in different states (s) (similar to the quantile results presented

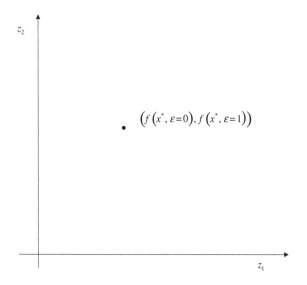

Fig. 11.2. Single space tradeoff for two random outcomes for a given controllable input vector.

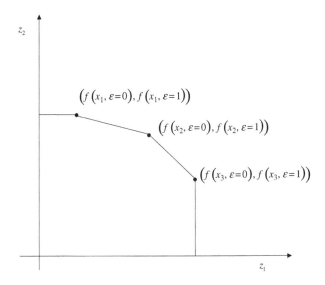

Fig. 11.3. State spaces possible for changing input levels.

in Table 11.1). In general, the set of possible states is defined as $\Omega = \{1, 2, \ldots, S\}$, the set of possible consequences is defined as $\Upsilon \subseteq R^M$, and actions are a mapping from the set of possible states to the set of consequences $A : \Omega \to R^M$. Each action defines a set of state contingent vectors of outcomes $y(A) \in R^{M \times S}$. It is important to note that the decision maker chooses the action based on its potential consequences for each state.

Finally, it is important to note that the amount of input applied is dependent on the output state. In the discussion of Chambers and Quiggin, this is referred to as a *state-allocable input technology*. In the case of the Bernoulli variable above, assume that $\varepsilon = 0$ denotes a drought while $\varepsilon = 1$ denotes adequate rainfall. In normal years, much of Midwestern agriculture removes excess water from fields using a system of tiles buried under the ground. However, drought condition expenditures on drainage may be replaced by expenditures on irrigation. Thus, fixing the total amount of money spent on drainage and irrigation, the marginal physical product for each allocation is depicted in Fig. 11.4. Specifically, $x_{1,2}$ is the amount of input expended in state $\varepsilon = 0$ (i.e., a normal rainfall year). However, in a high rainfall

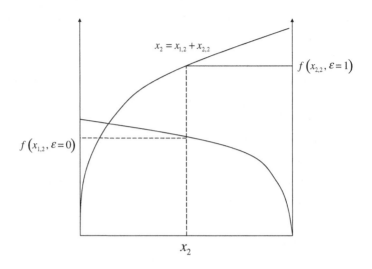

Fig. 11.4.   State contingent input decisions.

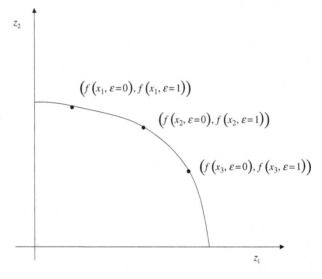

Fig. 11.5.   Shifting input across state-space.

year or state, $\varepsilon = 1$, the expenditure on drainage, increases to x_2 (or $x_2 = x_{1,2} + x_{2,2}$ where $x_{2,2}$ is applied in wet years). Reallocating between $x_{1,i}$ and $x_{2,i}$ yields a smooth tradeoff between the states as depicted in Fig. 11.5.

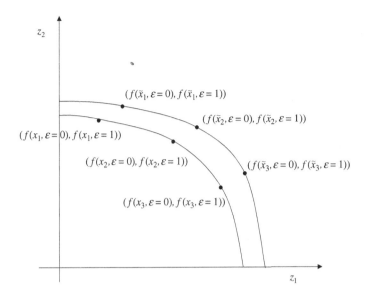

Fig. 11.6. Increasing input level across state-space.

Figure 11.6 depicts the increased level of total input from x to $\tilde{x}$. The outward shift represents the increased input level while the curves depict the reallocation of inputs between states. Note the effects of each decision. Choosing between $\{f(x_1, \varepsilon = 0), f(x_1, \varepsilon = 1)\}$ and $\{f(x_2, \varepsilon = 0), f(x_2, \varepsilon = 1)\}$ involves a choice between state contingent outcomes for a fixed total level of input x. On the other hand, the choice between $\{f(x_1, \varepsilon = 0), f(x_1, \varepsilon = 1)\}$ and $\{f(\tilde{x}_1, \varepsilon = 0), f(\tilde{x}_1, \varepsilon = 1)\}$ denotes the choice in the overall input level with a comparable commitment in each state.

11.2. State Production Set and Input Requirement Set

Based on this state-preference structure, Chambers and Quiggin develop a system similar to that of deterministic production analysis. The *state-contingent output correspondence* is the set of state-contingent outputs that can be generated from a given set of inputs:

$$Z(x) = \left\{ z \in R^{M \times X} : x \in R^N \text{ can produce } z \right\}. \tag{11.13}$$

The *input correspondence* is the set of inputs that can generate a possible set of states:

$$X(z) = \{x \in R_+^N : x \text{ can produce } z \in R^{M \times S}\}. \qquad (11.14)$$

This is the set of inputs that could be used to produce a state-contingent set of outputs. Note that $\{z_1(x_1), z_2(x_1)\}$ in Fig. 11.6 is a state-contingent set of outputs. If $\varepsilon = 0$, the outcome is $z_2(x_1)$ and if the state is $\varepsilon = 1$, then the outcome is $z_1(x_1)$. Based on the probabilities, each state will occur conditioned on the level of input x_1 with a certain probability. $X(z)$ in Eq. (11.14) is then that set of inputs that can be used to generate a given set of state-contingent outputs. Both the state-contingent output correspondence and the input correspondence obey properties similar to those properties governing the deterministic production function (Chambers, 1988).

11.3.   Distance Functions and Risk Aversion

The state-contingent output maps the set of feasible output states that can be generated from a vector of inputs. The next question is: What does this have to do with risk aversion or making input decisions? This section uses the state contingent mapping functions in Eqs. (11.13) and (11.14) to derive possible economic behavior in much the same way that level sets and production functions are used to construct the optimizing behavior in the dual production model.

11.3.1.   *Defining the Distance Function in State-Contingency Space*

One tool used in production analysis is the distance function. In production theory, the *distance function* is typically defined as the fraction of input that can be taken away in a way that still satisfies a level set (or level of production). Specifically,

$$D_I(x, z) = \max_{\theta} \left\{ \theta : \left(\frac{x}{\theta}\right) \in X(z) \right\}, \qquad (11.15)$$

where θ is a scalar distance to be determined by the formulation, x is a vector of inputs, and $X(z)$ is a level set (a set that produces at least output z as defined in Eq. (11.14)). Under typical assumptions, $\theta \leq 1$ (or the frontier is the best possible result similar to the derivations from duality). The distance function is graphically depicted in Fig. 11.7.

Using a similar justification, Chambers and Quiggin propose a *state-contingent output-distance function*

$$O(z, x) = \inf \left\{ \theta > 0 : \frac{z}{\theta} \in Z(x) \right\}. \tag{11.16}$$

The state-contingent output-distance function is presented graphically in Fig. 11.8. Using this definition, for $\theta < 1$, the point is on the interior of the state-contingent output function (again following the standard result from duality). From Fig. 11.5, note that the output frontier itself

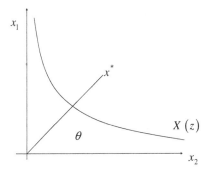

Fig. 11.7.   Distance function in input space.

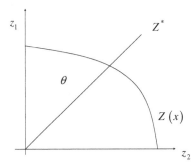

Fig. 11.8.   State contingent distance function.

is defined as

$$\bar{Z}(x) = \{z : O(z,x) = 1\}. \tag{11.17}$$

This function also determines the *state-contingent marginal rate of transformation* or the rate at which one state has to be traded for another. This transformation has something to say about choice of outputs across states. Specifically, the rate at which output in one state can be traded for output in another state can be formulated as

$$\frac{\partial z_2}{\partial z_1} = \frac{\frac{\partial O(z,x)}{\partial z_1}}{\frac{\partial O(z,x)}{\partial z_2}}. \tag{11.18}$$

11.3.2. *Risk Aversion and Valuing States*

In general terms, Chambers and Quiggin define a value-mapping function that aggregates the value of outcomes in each state. This general mapping function is denoted

$$W : \Upsilon^S \to R, \tag{11.19}$$

where W is a space of input prices, Υ is the state-contingent income space, and R is the univariate (dollar) space. Under risk neutrality, this mapping function simply becomes the expected value:

$$W(y) = \sum_{s \in \Omega} \pi_s y_s, \tag{11.20}$$

where π_s denotes the probability space s. Similarly under expected utility

$$W(y) = \sum_{s \in \Omega} \pi_s u(y_s), \tag{11.21}$$

where $u(y_s)$ is a utility function.

Backing up slightly, to define this value function, income can be used as the measure of well-being resulting from a given state of nature. In this case, income (y_s) is defined as

$$y_s = r_s - g(x), \tag{11.22}$$

where r_s is defined as the revenue in a particular state:

$$r_s(z_s) = \sum_m p_{ms} z_{ms}, \tag{11.23}$$

where p_{ms} denotes the output price for the m outputs in each state s (z_{ms}). Note that output levels, output prices, and input prices may vary across states of nature (i.e., each state of nature could have its own output price determined by the supply and demand conditions in that state).

11.3.3. *Duality, Benefit, and Distance Functions*

Following the distance function approach from duality, Chambers and Quiggin develop the *benefit function* based on general value mapping:

$$B(w, y) = \max\{\beta \in R : W(y - \beta g) \geq w\} \tag{11.24}$$

if $W(y - \beta g) \geq w$ for some β. Following Eq. (11.18), Υ^* is a specific set of states (as defined by the mapping function in Eq. (11.19)). Υ^* is above $W(y) = w$ for a specific input cost. Thus, β is the constant amount in the direction g that can be subtracted before the state-space is on $W(y)$. The benefit function is graphically depicted in Fig. 11.9 For example, assuming risk neutrality:

$$W(y) = \sum_{s \in \Omega} \pi_s(y_s - \beta g). \tag{11.25}$$

The point above the output frontier is the observed benefit (remember that the axes are basically incomes in each state). The benefit function is then the vertical shift down to the level set (a fixed value of the valuation function).

Properties of the benefit function:

- $B(w, y)$ is nonincreasing in w and nondecreasing in y
- $B(w, y + \alpha 1) = B(w, y) + \alpha, \alpha \in R$
- $B(w, y) \geq 0 \Leftrightarrow W(y) \geq w$ and $B(w, y) = 0 \Leftrightarrow W(y) = w$

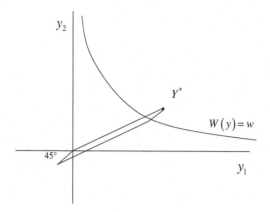

Fig. 11.9.   Benefit function.

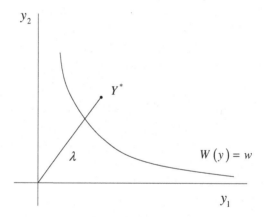

Fig. 11.10.   Distance function formulation of the state benefit.

A second formulation is the Malmquist distance function

$$D(y, w) = \sup\left\{\lambda > 0; W\left(\frac{y}{\lambda}\right) \geq w\right\}, \qquad (11.26)$$

which is the flip side of the distance function discussed above, as presented in Fig. 11.10.

To make the functions from Eqs. (11.24) and (11.26) a little more transparent, assume that the decision maker is applying inputs associated with Υ^* in Figs. 11.9 and 11.10. Assuming that the desired state-contingent point lies on the surface of $W(y)$, how much input could be taken away from production? This has two answers: in Eq. (11.24), the quantity of inputs is βg and in Eq. (11.26), the fraction inputs is λ.

Building on this formulation, the certainty equivalent is the smallest certain income that leaves the individual indifferent to the risky gamble. This can be defined based on the state-contingent model as

$$e(y) = \inf\{c \in R : W(c1) \geq W(y)\}. \qquad (11.27)$$

This function can be represented as a combination of the Malmquist distance function and the benefit function:

$$e(y) = \frac{1}{D(1, W(y))} = -B(W(y), 0). \qquad (11.28)$$

11.3.4. *Defining Risk Aversion Graphically*

Figure 11.11 presents a slightly different depiction of the state-space developed in the preceding sections. In this state-space graph, the 45° line is essentially a risk-free asset (on this line the income is the same for each state). Alternatively, all returns on the *fair-odds line* have the same expected value. From the producer's point of view, $-\pi_1/\pi_2$ denotes the fair-odds, or that locus of points that have the same expected value. Only at the point where the fair-odds line crosses the 45° line is the decision made by the producer risk-free.

As a result, the decision maker is *risk averse* with respect to the probability vector π if

$$W(\bar{y}1) \geq W(y) \ \forall y \qquad (11.29)$$

(Chambers and Quiggin, 2000, Definition 3.1, p. 88). Indifference implies a convex set of preferences around point A. Put differently, the payoffs must be "better than fair" to make an individual indifferent between the risk-less payoff and the certain payoff of the 45° line.

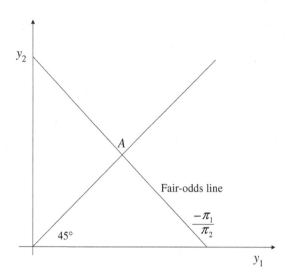

Fig. 11.11.   Risk-free states and fair-odds line.

11.3.5.   *Constant Relative and Absolute Risk Aversion*

The differences between the fair-odds line and the tradeoff actually made by decision makers can be used to derive the degree of risk aversion. However, these derivations are somewhat different than the values derived in Chapter 4.

11.3.5.1.   *Risk Premium*

Based on the definition of risk aversion from the preceding section, Chambers and Quiggin define the *risk premium* as the amount that a risk averse decision maker is willing to sacrifice to obtain a certain payoff. In Fig. 11.12, y is an outcome from a *state-contingent output combination*, however given risk aversion, the decision maker is indifferent between the state-contingent output combination y^* and y (the distance between A and y^* on the 45° line). Note that y^* is a certain payoff because it is on the 45° line. The certainty equivalent is the distance between point y and point y^* in state-space. Analytically, the

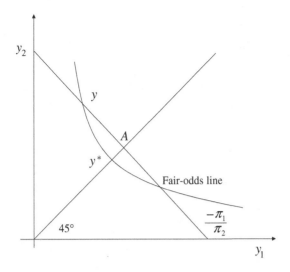

Fig. 11.12. Defining risk aversion graphically.

absolute risk premium ($r(y)$) is defined by the benefit function:

$$r(y) = \max\{c : W((\bar{y} - c)1) \geq W(y)\}$$
$$= B(W(y), \bar{y}1). \tag{11.30}$$

The *relative risk premium* ($v(y)$) is then defined by the distance function:

$$v(y) = \sup\left\{\lambda > 0 : W\left(\frac{\bar{y}1}{\lambda}\right) \geq W(y)\right\}. \tag{11.31}$$

Mathematically constant absolute risk aversion can be defined as:

Definition 3.2 in Chambers and Quiggin, 2000, p. 97: W displays constant absolute risk aversion if, for any y, for $t \in R$.

$$r(y + t1) = r(y). \tag{11.32}$$

This definition states that the absolute risk premium remains unchanged if a constant outcome is added to each state (as demonstrated in Fig. 11.13).

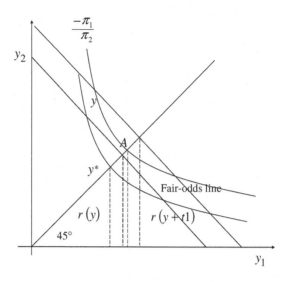

Fig. 11.13.   Using a constant shift to derive risk aversion.

Constant relative risk aversion can then be defined as

Definition 3.3 in Chambers and Quiggin, 2000, p. 97: W displays constant relative risk aversion if, for any y, for $t \in R$

$$v(ty) = v(y). \tag{11.33}$$

11.3.5.2.   *Derivation of the Effort Function*

Returning to the notion of the valuation function from Eq. (11.22)

$$y_s = r_s - g(x)$$
$$r_s(z_s) = \sum_m p_{ms} z_{ms} \tag{11.34}$$
$$W(y) = \sum_{s \in \Omega} \pi_s u(y_s).$$

From this formulation, the valuation is dependent on the amount of effort applied (inputs used). This concept can be used to define the *effort–cost functions*. The effort function yields what the input bundle is worth to the decision maker (i.e., the state valuation associated with Eq. (11.34)).

Properties of the Effort–Evaluation Function

- g is nondecreasing and continuous for all $x \in R$.
- $g(\mu x) = \mu g(x)$ for all $\mu > 0$ and $x \in R$.
- $g(x + x^0) \leq g(x) + g(x^0)$ for all $x, x^0 \in R$.

Based on these properties, Chambers and Quiggin define the effort–cost function as

$$c(z) = \min_x \{g(x) : x \in X(z)\}. \tag{11.35}$$

Note that this result states that the effort–cost function is defined as the minimum cost method of producing a given set of state contingent outputs.

Remember that $X(z)$ is the set of all inputs that can be used to produce a vector of state contingent outputs, z. Figure 11.14 depicts this input requirement set from the state-contingency space. Following standard production economics, the optimum is determined where the convex surface is tangent to the input price ratio.

If $g(x)$ is a linear function, this relationship looks a lot like the standard cost-minimization model from duality. In fact, if the effort function is linear in input prices:

$$c(w, z) = \min_x \{w'x : x \in X(z)\} \tag{11.36}$$

where w is the vector of input prices. However, note that z is not a vector of deterministic outputs, but a vector of state-contingent outputs.

Shephard's Lemma in Chambers and Quiggin, 2000, p. 127: If a unique solution, $x(w, z) \in R$, exists to the minimization problem, the cost function is differentiable in w and

$$x(w, z) = \nabla_w c(w, z) \tag{11.37}$$

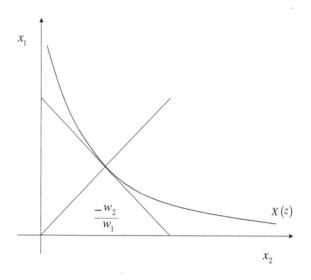

Fig. 11.14.   Optimal input choice based on state-contingent outputs.

and, if the cost function is differentiable in w, there exists a unique solution to the cost minimization problem and

$$\nabla_w c(w, z) = x(w, z). \qquad (11.38)$$

Putting it together, under risk neutrality, the decision maker chooses the state-contingent outputs to maximize the expected return on production:

$$\max_x \left\{ \sum_{s \in \Omega} \pi_s \sum_{m=1}^{M} p_{ms} z_{ms} - c(w, z) \right\}, \qquad (11.39)$$

which can be transformed to

$$\max_r \left\{ \sum_{s \in \Omega} \pi_s r_s - C(w, r, p) \right\}. \qquad (11.40)$$

Here $C(w, r, p)$ is the *revenue–cost function* defined as

$$C(w, r, p) = \min \left\{ c(w, z) : \sum_{m=1}^{M} p_{ms} z_{ms} \geq r_s, s \in \Omega \right\}. \qquad (11.41)$$

To demonstrate the implications of this model, consider the formulation under constant absolute riskiness.

First, Chambers and Quiggin define the *certainty equivalent revenue* as

$$e^c(r, p) = \sup\{e : C(e1, p) \leq C(r, p)\} \qquad (11.42)$$

defining the revenue vector as

$$T(r, p) = e^c(r, p). \qquad (11.43)$$

Lemma 5.2 in Chambers and Quiggin, 2000, p. 169: The technology displays constant absolute riskiness if and only if the revenue–cost function can be represented as

$$C(w, r, p) = \hat{C}(w, T(r, p, w), p), \qquad (11.44)$$

where

$$T(r + \delta 1, p, w) = T(r, p, w) + \delta$$

$$T(\lambda r, \lambda p, \lambda w) = \lambda T(r, p, w), T(r, p, \lambda w) = T(r, p, w). \qquad (11.45)$$

11.4.   Summary

This book has presented a variety of risk models where the randomness was simply appended onto a production relationship. As a result, there was little interaction between the choice of input levels and risk. This chapter introduced two procedures for integrating this relationship between risk and input choices. The stochastic production frontier builds on previous work by Just and Pope (1978). The empirical techniques are fairly feasible. This chapter presents a quantile regression approach, which is somewhat different from the traditional approach. The second approach is based on state-contingency formulations. This chapter presents a synthesis of the model developed by Chambers and Quiggin.

Chapter 12

Risk, Uncertainty, and the Agricultural Firm — A Summary and Outlook

This text started by noting that economics is the study of choices. In this case, these choices were dependent on some factors that were outside the decision maker's control and largely unknowable at the time that most decisions were made (although Chapter 8 presents models where the decision could be modified according to new information). Given this scenario, the text draws on economic and statistical theory to demonstrate how consistent (or at least well-reasoned) decisions can be made. However, it is also noted that the best-reasoned decision cannot guarantee a good outcome.

With this overview in mind, the book begins the development of tools for decision making under risk with a presentation of statistics. Specifically, Chapter 2 presents the basic definition and an overview of various representations of random variables. Given the focus of this book, the emphasis is on the estimation of the parameters of the distribution functions instead of traditional emphasis on hypothesis testing.

Chapters 3 and 4 present the theoretical context for decision making under risk: the expected utility hypothesis and risk aversion. Chapter 3 provides a detailed description of the development of the expected utility hypothesis and how the expected utility hypothesis is linked to more frequently developed concepts from demand theory. While the text provides an intuitive description of the proof the expected utility hypothesis, an extremely detailed description of the proof is presented in Appendix D. In addition, Chapter 3 links the expected utility model to the expected value–variance formulation that is prevalent in most portfolio and capital market studies.

The mathematical mechanics of portfolio analysis is presented in Chapter 5. This development looks ahead to several formulations of the whole farm-planning model presented in Chapter 6 and the market models of risk presented in Chapter 10.

Chapter 7 presents a fairly rigorous treatment of risk efficiency criteria in the guise of stochastic dominance. Further, the chapter presents the development of stochastic dominance with respect to a function and some recent literature on the empirical implementation of stochastic dominance using nonparametric regression.

Chapter 10 presents several variants of derivatives including the traditional view of futures markets and option pricing models. In addition, the chapter examines the use of real option pricing to value investments under risk and uncertainty. Chapter 10 presents the notion of crop insurance as an option problem.

Chapter 11 presents an emerging area in the analysis of risk: the state-contingent formulation. This approach harkens back to the stochastic production function formulation to analyze input decisions as a part of the analysis of risk.

This book is intended to provide the student with a starting point for each of the methodologies. It is not intended to be a detailed discussion. Hence, it provides a link to the literature for the further development of each topic.

It seems that risk is a cyclical or sporadic topic of interest to the economics and agricultural economics question. The author's assumption is that risk is an important component of modeling the actions of the agricultural firm. The effectiveness of agricultural policy and well-being of agricultural producers and rural communities are intimately linked to farmers' response (or ability to respond) to risk and uncertainty.

However, there are empirical indications that farmers' response to risk is more complicated than the models presented in this text would indicate. Notably, farmers participate in insurance programs (which are in part subsidized by the federal government in the United States) on a limited basis. Thus, either they possess other mechanisms to control risk, they value firm growth more than risk reduction, or assume that the sector can successfully extract rents from the federal government through other programs (i.e., disaster payments).

Appendix A

Measure Theory and the Justification
of Random Variables

A measure is an extended real-valued, non-negative, and countably additive set function μ, defined on a ring $\mathbf{R}$, and such that $\mu(\emptyset) = 0$. Figure A.1 presents a ring of points on the real number line.

Assume the set mapping function from these sets of xs to the real number line

$$\mu(\{x\}) = \frac{1}{2}(x_j - x_i)^2$$

$$\mu : X \to R_1.$$

(A.1)

Consider whether the set function defined in Eq. (A.1) is consistent with a measure for a series of subintervals defined over the interval of the real number line between 1 and 5. Starting with unitary intervals

$$E_1 = [x_1, x_2) = [1, 2) \Rightarrow \mu(E_1) = \frac{1}{2}$$

$$E_2 = [2, 3) \Rightarrow \mu(E_2) = \frac{1}{2}$$

$$E_3 = [3, 4) \Rightarrow \mu(E_3) = \frac{1}{2}$$

$$E_4 = [4, 5) \Rightarrow \mu(E_4) = \frac{1}{2}.$$

(A.2)

Fig. A.1.   Ring defined on a number line.

Reconstructing the set function on two-unit intervals

$$E_1^* = [1,3) \Rightarrow \mu(E_1^*) = 2$$
$$E_1^* = [3,5) \Rightarrow \mu(E_2^*) = 2.$$
(A.3)

And, finally reconstructing the set function on a four-unit interval

$$E_1^{**} = [1,5) \Rightarrow \mu(E_1^{**}) = 8.$$
(A.4)

Can this set function be a measure? The function is a positive real-valued function. However, it is not countably additive

$$\mu(E_1^{**}) > \mu(E_1^*) + \mu(E_2^*) > \mu(E_1) + \mu(E_2) + \mu(E_3) + \mu(E_4)$$
$$8 > 2 + 2 > \frac{1}{2} + \frac{1}{2} + \frac{1}{2} + \frac{1}{2}.$$
(A.5)

Hence, the set-valued function defined in Eq. (A.1) is not a measure. Consider the trivial case of measure

$$\mu(\{x\}) = |x_j - x_i|$$
(A.6)

or the standard distance formula

$$\mu^*(\{x\}) = [(x_j - x_i)^2]^{\frac{1}{2}}.$$
(A.7)

Applying the standard distance function to each interval starting with the unit intervals defined in Eq. (A.2) yielding

$$E_1 = [1,2) \Rightarrow \mu^*(E_1) = 1$$
$$E_2 = [2,3) \Rightarrow \mu^*(E_2) = 1$$
$$E_3 = [3,4) \Rightarrow \mu^*(E_3) = 1$$
$$E_4 = [4,5) \Rightarrow \mu^*(E_4) = 1.$$
(A.8)

Next calculate the distance function for the two-unit intervals defined in Eq. (A.3)

$$E_1^* = [1,3) \Rightarrow \mu^*(E_1^*) = 2$$
$$E_1^* = [3,5) \Rightarrow \mu^*(E_2^*) = 2. \tag{A.9}$$

Finally, calculate the distance function for the total interval

$$E_1^{**} = [1,5) \Rightarrow \mu^*(E_1^{**}) = 4. \tag{A.10}$$

Thus, it is concluded that the standard distance function is consistent with the additivity requirement for a measure:

$$\mu^*(E_1^{**}) = \mu^*(E_1^*) + \mu^*(E_2^*) = \mu^*(E_1) + \mu^*(E_2) + \mu^*(E_3) + \mu^*(E_4)$$

$$4 > 2 + 2 > 1 + 1 + 1 + 1. \tag{A.11}$$

Noting the correspondence with the probability function: Let $P[\{x\}]$ be a measure function defined on a portion of the real number line (i.e., $x \in [a, b]$). The measure function on this interval is then

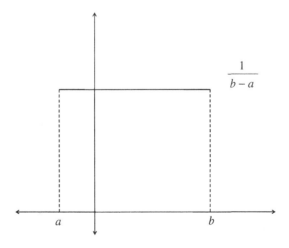

Fig. A.2. The uniform distribution as a set-mapping function.

defined as

$$P[[x_i, x_j)] = \int_{\max(x_i,a)}^{\min(x_j,b)} \frac{1}{b-a} dx = \frac{1}{b-a}(\min(x_j, b) - \max(x_i, a)).$$

$$(A.12)$$

This is a real-valued set function, whose result is positive. Graphically, the countably additive characteristics can be demonstrated in Fig. A.2.

Appendix B

Derivation of the Moments of the Inverse Hyperbolic Sine Distribution

The inverse hyperbolic sine transformation to normality used by Moss and Shonkwiler (1993) is a redefinition of Burbidge, Magee, and Robb (1988) transformation as modified by Ramirez (1992). The difference is that Burbidge, Magee, and Robb transform the residual from a regression rather than the random variable itself. Ramirez reformulates the transformation as

$$\frac{\ln(\theta[\, y_t - x_t\beta] + [\theta^2\{y_t - x_t\beta\}^2 + 1]^{\frac{1}{2}})}{\theta} \sim N(\delta, \sigma^2) \qquad (\text{B.1})$$

where θ is a parameter of transformation, y_t is the observed level of the dependent variable in period t, x_t is the observed level of the independent variables, β are the estimated values of the regression, δ is the mean of the transformed variable, and σ^2 is the variance of the transformed variable. For simplification purposes, abstract away from the time series formulation

$$\frac{\ln(\theta[\, y - \mu] + [\theta^2\{y - \mu\}^2 + 1]^{\frac{1}{2}})}{\theta} \sim N(\delta, \sigma^2) \qquad (\text{B.2})$$

where y is a random variable and μ is another parameter of the distribution. Writing this expression as a trigonometric function

$$\nu = \frac{\sinh^{-1}(\theta[y - \mu])}{\theta} \Leftarrow \sinh^{-1}(z) = \ln(z + [1 + z^2]^{\frac{1}{2}}) \qquad (\text{B.3})$$

257

where $\sinh^{-1}(z)$ denotes the inverse hyperbolic sine transformation sometimes written as $\text{arcsinh}(z)$. Next, define the hyperbolic sine transformation as

$$z = \frac{\sinh(\theta y)}{\theta} = \frac{\exp[\theta y] - \exp[-\theta y]}{2\theta} \Leftarrow \sinh(v) = \frac{\exp[z] - \exp[-z]}{2}$$

(B.4)

where $\sinh(y)$ denotes the hyperbolic sine transformation.

The inverse hyperbolic sine transformation function to normality can then be written as

$$y = \phi(x) \Rightarrow y - \mu = \frac{\sinh(\theta x)}{\theta}.$$

(B.5)

Defining x in terms of y

$$x = \phi^{-1}(y) \Rightarrow y - \mu = \frac{\sinh(\theta x)}{\theta}$$

$$\Rightarrow \theta[y - \mu] = \sinh(\theta x)$$

$$\Rightarrow \sinh^{-1}(\theta[y - \mu]) = \theta x$$

(B.6)

$$\Rightarrow \frac{\sinh^{-1}(\theta[y - \mu])}{\theta} = x, x \sim N(\delta, \sigma^2).$$

The probability density function of y then becomes

$$g(y) = f\left(\frac{\sinh^{-1}(\theta[y - \mu])}{\theta}\right) \left| \frac{d}{dy} \frac{\sinh^{-1}(\theta[y - \mu])}{\theta} \right|.$$

(B.7)

The Jacobian term

$$\frac{d}{dy} \frac{\sinh^{-1}(\theta[y - \mu])}{\theta}$$

$$= \frac{d}{dy} \frac{\ln(\theta[y - \mu] + [\theta^2\{y - \mu\}^2 + 1]^{\frac{1}{2}})}{\theta}$$

$$= \frac{(\theta + \frac{1}{2}[\theta^2\{y - \mu\}^2 + 1]^{\frac{1}{2}}[2\theta^2\{y - \mu\}])}{(\theta[y - \mu] + [\theta^2\{y - \mu\}^2 + 1]^{\frac{1}{2}})\theta} \times \frac{(\theta^2\{y - \mu\}^2 + 1)^{\frac{1}{2}}}{(\theta^2\{y - \mu\}^2 + 1)^{\frac{1}{2}}}$$

$$= \frac{(\theta[\theta^2\{y-\mu\}^2 + 1]^{\frac{1}{2}} + \theta^2\{y-\mu\})}{(\theta^2\{y-\mu\}^2 + 1)^{\frac{1}{2}}(\theta[y-\mu] + [\theta^2\{y-\mu\}^2 + 1]^{\frac{1}{2}})\theta}$$

$$= \frac{\theta([\theta^2\{y-\mu\}^2 + 1]^{\frac{1}{2}} + \theta\{y-\mu\})}{(\theta^2\{y-\mu\}^2 + 1)^{\frac{1}{2}}(\theta[y-\mu] + [\theta^2\{y-\mu\}^2 + 1]^{\frac{1}{2}})\theta}$$

$$= \frac{1}{(\theta^2\{y-\mu\}^2 + 1)^{\frac{1}{2}}} = (\theta^2\{y-\mu\}^2 + 1)^{\frac{1}{2}}. \tag{B.8}$$

Therefore, the probability density function for the transformed random variable becomes

$$g(y|\theta, \mu, \delta, \sigma^2) = \frac{1}{\sqrt{2\pi\sigma^2}} \exp\left[-\frac{1}{2\sigma^2}\left\{\left(\frac{\sinh^{-1}(\theta[y-\mu])}{\theta}\right) - \delta\right\}^2\right]$$

$$\times (\theta^2\{y-\mu\}^2 + 1)^{\frac{1}{2}}. \tag{B.9}$$

While Eq. (B.9) could be integrated with over y, this integral is overly complicated. Instead of integrating over y, integrate over the transformation of y back into x. Specifically, from Eq. (B.4) integrate

$$\mu_r(\varepsilon) = \frac{1}{\sqrt{2\pi\sigma^2}} \int_{-\infty}^{\infty} \left(\frac{\exp[\varepsilon\theta] - \exp[-\varepsilon\theta]}{2\theta}\right)^r \exp\left(-\frac{1}{2}\frac{[\varepsilon - \delta]^2}{\sigma^2}\right) d\varepsilon. \tag{B.10}$$

Letting $r = 1$ the expectation in Eq. (B.10) defines the mean as

$$\mu_1(\varepsilon) = \frac{1}{\sqrt{2\pi\sigma^2}} \int_{-\infty}^{\infty} \left(\frac{\exp[\varepsilon\theta] - \exp[-\varepsilon\theta]}{2\theta}\right) \exp\left(-\frac{1}{2}\frac{[\varepsilon - \delta]^2}{\sigma^2}\right) d\varepsilon$$

$$= \frac{1}{2\theta\sqrt{2\pi\sigma^2}}$$

$$\times \int_{-\infty}^{\infty} \left(\exp\left[\varepsilon\theta - \frac{(\varepsilon - \delta)^2}{2\sigma^2}\right] - \exp\left[-\varepsilon\theta - \frac{(\varepsilon - \delta)^2}{2\sigma^2}\right]\right) d\varepsilon. \tag{B.11}$$

Expanding the first term in the integral yields

$$\exp\left[-\frac{(\varepsilon-\delta)^2}{2\sigma^2}-\theta\varepsilon\right]=\exp\left[\frac{1}{2\sigma^2}(\varepsilon^2-2\delta\varepsilon+\delta^2-2\sigma^2\theta\varepsilon)\right]$$

$$=\exp\left[\frac{1}{2\sigma^2}(\varepsilon^2-2\varepsilon\{\delta+\theta\sigma^2\}+\delta^2)\right].$$

$$(B.12)$$

Completing the square in the exponent term

$$(\delta+\theta\sigma^2)^2=\delta^2+2\delta\theta\sigma^2+\theta^2(\sigma^2)^2 \qquad (B.13)$$

which implies that $+2\delta\theta\sigma^2+\theta^2(\sigma^2)^2-2\delta\theta\sigma^2-\theta^2(\sigma^2)^2=0$ can be added to the exponent yielding

$$\exp\left[-\frac{(\varepsilon-\delta)^2}{2\sigma^2}-\theta\varepsilon\right]=\exp\left[-\frac{(\varepsilon-(\delta-\theta\sigma^2))^2}{2\sigma^2}+\delta\theta+\frac{\theta^2\sigma^2}{2}\right].$$

$$(B.14)$$

Similarly, the second exponential under the integral becomes

$$\exp\left[-\frac{(\varepsilon-\delta)^2}{2\sigma^2}-\varepsilon\theta\right]=\exp\left[-\frac{(\varepsilon-\{\delta-\theta\sigma^2\})^2}{2\sigma^2}-\delta\theta+\frac{\theta^2\delta^2}{2}\right].$$

$$(B.15)$$

The complete integral from Eq. (B.11) then becomes

$$\mu_1(\varepsilon)=\frac{1}{2\theta}\exp\left(\delta\theta+\frac{\theta^2\sigma^2}{2}\right)\frac{1}{\sqrt{2\pi\sigma^2}}$$

$$\times\int_{-\infty}^{\infty}\exp\left(-\frac{[\varepsilon-\{\delta+\theta\sigma^2\}]^2}{2\sigma^2}\right)d\varepsilon$$

$$-\frac{1}{2\theta}\exp\left(-\delta\theta+\frac{\theta^2\sigma^2}{2}\right)\frac{1}{\sqrt{2\pi\sigma^2}}$$

$$\times\int_{-\infty}^{\infty}\exp\left(-\frac{[\varepsilon-\{\delta-\theta\sigma^2\}]^2}{2\sigma^2}\right)d\varepsilon. \qquad (B.16)$$

Noting that the two integrals in Eq. (B.16) are simply normal distributions with different means (see the development of the moment generating function of the normal distribution in Eq. (2.78)) the expectation becomes

$$\mu_1(\varepsilon) = \frac{1}{2\theta} \exp\left(\frac{\theta^2 \sigma^2}{2}\right) (\exp[\delta\theta] - \exp[-\delta\theta]). \qquad (B.17)$$

Working with the second moment, define $r = 2$ in Eq. (B.10) yielding

$$\mu_2(\varepsilon) = \frac{1}{\sqrt{2\pi\sigma^2}} \int_{-\infty}^{\infty} \left(\frac{\exp[\varepsilon\theta] - \exp[-\varepsilon\theta]}{2\theta}\right)^2 \exp\left(-\frac{1}{2}\frac{[\varepsilon - \delta]^2}{\sigma^2}\right) d\varepsilon. \qquad (B.18)$$

Expanding the first term under the integral

$$\left(\frac{\exp[\varepsilon\theta] - \exp[-\varepsilon\theta]}{2\theta}\right)^2$$

$$= \frac{\exp(\varepsilon\theta)\exp(\varepsilon\theta) - 2\exp(\varepsilon\theta)\exp(-\varepsilon\theta) + \exp(-\varepsilon\theta)\exp(-\varepsilon\theta)}{4\theta^2}$$

$$= \frac{\exp(\varepsilon\theta + \varepsilon\theta) - 2\exp(\varepsilon\theta - \varepsilon\theta) + \exp(-\varepsilon\theta - \varepsilon\theta)}{4\theta^2}$$

$$= \frac{\exp(2\varepsilon\theta) + \exp(-2\varepsilon\theta) - 2}{4\theta^2}. \qquad (B.19)$$

Thus, the integral becomes

$$\mu_2(\varepsilon) = \frac{1}{\sqrt{2\pi\sigma^2}} \int_{-\infty}^{\infty} \left(\frac{\exp(2\varepsilon\theta) + \exp(-2\varepsilon\theta) - 2}{4\theta^2}\right)$$

$$\times \exp\left(-\frac{1}{2}\frac{[\varepsilon - \delta]^2}{\sigma^2}\right) d\varepsilon$$

$$= \frac{1}{4\theta^2\sqrt{2\pi\sigma^2}} \int_{-\infty}^{\infty}$$

$$\times \left(\exp\left[-\frac{[\varepsilon - \delta]^2}{2\sigma^2} + 2\varepsilon\theta\right] + \exp\left[-\frac{[\varepsilon - \delta]^2}{2\sigma^2} - 2\varepsilon\theta\right] - 2\right) d\varepsilon. \qquad (B.20)$$

Decomposing the integral using the summation, the first term can be simplified as

$$\frac{1}{4\theta^2\sqrt{2\pi\sigma^2}}\int_{-\infty}^{\infty}\exp\left[-\frac{(\varepsilon^2-2\varepsilon\delta+\delta^2)}{2\sigma^2}+2\varepsilon\theta\right]d\varepsilon$$

$$=\frac{1}{4\theta^2\sqrt{2\pi\sigma^2}}\int_{-\infty}^{\infty}\exp\left[\frac{-\varepsilon^2+2\varepsilon\delta-\delta^2+4\varepsilon\theta\sigma^2}{2\sigma^2}\right]d\varepsilon$$

$$=\frac{1}{4\theta^2\sqrt{2\pi\sigma^2}}\int_{-\infty}^{\infty}\exp\left[\frac{-\varepsilon^2+2\varepsilon(\delta+2\theta\sigma^2)-\delta^2}{2\sigma^2}\right]d\varepsilon$$

$$=\frac{1}{4\theta^2\sqrt{2\pi\sigma^2}}\int_{-\infty}^{\infty}\exp\left[-\frac{(\varepsilon-(\delta+2\theta\sigma^2))^2}{2\sigma^2}+2\delta\theta+2\theta^2\sigma^2\right]d\varepsilon$$

$$=\frac{1}{4\theta^2}\exp[2\delta\theta+2\theta^2\sigma^2]. \tag{B.21}$$

By a similar expansion of the second term under the integral of Eq. (B.20) and recognizing the third term as a constant yields

$$\mu_2(\varepsilon)=\frac{1}{4\theta^2}(\exp[2\delta\theta+2\theta^2\sigma^2]+\exp[-2\delta\theta+2\theta^2\sigma^2]-2). \tag{B.22}$$

Extending this expression to the third moment

$$\mu_3(\varepsilon)=\frac{1}{\sqrt{2\pi\sigma^2}}\int_{-\infty}^{\infty}\left(\frac{\exp(\theta\varepsilon)-\exp(-\theta\varepsilon)}{2\theta}\right)^3\exp\left(-\frac{1}{2}\frac{(\varepsilon-\delta)^2}{\sigma^2}\right)d\varepsilon$$

$$=\frac{1}{8\theta^3\sqrt{2\pi\sigma^2}}\int_{-\infty}^{\infty}(\exp(3\theta\varepsilon)-3\exp(\theta\varepsilon)+3\exp(-\theta\varepsilon)$$

$$-\exp(-3\theta\varepsilon))\exp\left(-\frac{1}{2}\frac{(\varepsilon-\delta)^2}{\sigma^2}\right)d\varepsilon. \tag{B.23}$$

Following the procedure used above, start by working the first term under the integral out as

$$\exp\left(-\frac{(\varepsilon-\delta)^2}{2\sigma^2}+3\theta\varepsilon\right)$$

$$=\exp\left(\frac{-\varepsilon^2+2\varepsilon\delta-\delta^2+6\theta\varepsilon\sigma^2}{2\sigma^2}\right)$$

$$= \exp\left(\frac{-\varepsilon^2 + 2\varepsilon[\delta + 3\theta\sigma^2] - \delta^2}{2\sigma^2}\right)$$

$$= \exp\left(\frac{\begin{array}{c}-\varepsilon^2 + 2\varepsilon[\delta + 3\theta\sigma^2] - \delta^2 - 6\delta\theta\sigma^2 \\ -9\theta^2(\sigma^2)^2 + 6\delta\theta\sigma^2 + 9\theta^2(\sigma^2)^2\end{array}}{2\sigma^2}\right)$$

$$= \exp\left(\frac{-\varepsilon^2 + 2\varepsilon[\delta + 3\theta\sigma^2] - (\delta + 3\theta\sigma^2)^2 + 6\delta\theta\sigma^2 + 9\theta^2(\sigma^2)^2}{2\sigma^2}\right)$$

$$= \exp\left(\frac{-(\varepsilon - (\delta + 3\theta\sigma^2))^2}{2\sigma^2} + 3\delta\theta + \frac{9}{2}\theta^2\sigma^2\right). \tag{B.24}$$

Following similar transformations

$$-3\exp[\theta\varepsilon]\exp\left[-\frac{(\varepsilon - \delta)^2}{2\sigma^2}\right]$$

$$= -3\exp\left[-\frac{(\varepsilon - (\delta + \theta\sigma^2))^2}{2\sigma^2} + \delta\theta + \frac{1}{2}\theta^2\sigma^2\right]$$

$$3\exp[-\theta\varepsilon]\exp\left[-\frac{(\varepsilon - \delta)^2}{2\sigma^2}\right]$$

$$= 3\exp\left[-\frac{(\varepsilon - (\delta - \theta\sigma^2))^2}{2\sigma^2} - \delta\theta + \frac{1}{2}\theta^2\sigma^2\right] \tag{B.25}$$

$$-\exp[-3\theta\varepsilon]\exp\left[-\frac{(\varepsilon - \delta)^2}{2\sigma^2}\right]$$

$$= -\exp\left[-\frac{(\varepsilon - 3\theta\sigma^2)^2}{2\sigma^2} - 3\delta\theta + \frac{9}{2}\theta^2\sigma^2\right].$$

The integral then becomes

$$\mu_3(\varepsilon) = \frac{1}{8\theta^3}\left(\exp\left[3\delta\theta + \frac{9}{2}\theta^2\sigma^2\right] - 3\exp\left[\delta\theta + \frac{1}{2}\theta^2\sigma^2\right]\right.$$

$$\left. + 3\exp\left[-\delta\theta + \frac{1}{2}\theta^2\sigma^2\right] - \exp\left[-3\delta\theta + \frac{9}{2}\theta^2\sigma^2\right]\right). \tag{B.26}$$

The fourth moment of the distribution then becomes

$$\mu_4(\varepsilon) = \frac{1}{\sqrt{2\pi\sigma^2}} \int_{-\infty}^{\infty} \left(\frac{\exp(\theta\varepsilon) - \exp(-\theta\varepsilon)}{2\theta} \right)^4 \exp\left(-\frac{(\varepsilon - \delta)^2}{2\sigma^2} \right) d\varepsilon$$

$$= \frac{1}{16\theta^4 \sqrt{2\pi\sigma^2}} \int_{-\infty}^{\infty} (\exp[4\theta\varepsilon] - 4\exp[2\theta\varepsilon]$$

$$- 4\exp[-2\theta\varepsilon] + \exp[-4\theta\varepsilon] + 6)$$

$$\times \exp\left(-\frac{(\varepsilon-\delta)^2}{2\sigma^2} \right) d\varepsilon. \tag{B.27}$$

Following through on the integration as in Eq. (B.25)

$$\mu_4(\varepsilon) = \frac{1}{16\theta^4} (\exp(4\theta\delta + 8\theta^2\sigma^2) - 4\exp(2\theta\delta + 2\theta^2\sigma^2)$$

$$- 4\exp(-2\theta\delta + 2\theta^2\sigma^2) + \exp(-4\theta\delta + 8\theta^2\sigma^2) + 6). \tag{B.28}$$

Hence, the first four moments of the inverse hyperbolic sine distribution can be expressed as "closed-form" equations.

Appendix C

Numerical Techniques for Applied Optimization and Solution of Nonlinear Systems of Equations

While closed form optimums may exist for several of the estimation problems implied by the method of moments and maximum likelihood approach described above, several specifications such as the inverse hyperbolic sine specification cannot be readily solved analytically. One alternative used by studies such as Moss and Shonkwiler is the use of numerical optimization techniques. Most of these numerical techniques are based on modifications of the Newton–Raphson procedure. This appendix briefly presents the derivation of the classical optimality conditions and the Newton–Raphson procedure for numerical optimization.

Most students have been introduced to the first-order conditions for optimality. For the purpose of this text, this appendix redevelops these conditions within the framework of a second-order Taylor series expansion of a function. Specifically, taking the second-order Taylor series expansion of a function $f(x)$ of a vector $x \in M_{n \times 1}$ (or x is an $n \times 1$ vector)

$$f(x) = f(x_0) + \nabla_x f(x)(x - x_0)$$
$$+ \frac{1}{2}(x - x_0)' \nabla_{xx}^2 f(x)(x - x_0) + \varepsilon(x - x_0) \qquad \text{(C.1)}$$

where x_0 is the point of approximation, $\nabla_x f(x_0)$ is the gradient vector

$$\nabla_x f(x_0) = \left[\left. \frac{\partial f(x)}{\partial x_1} \right|_{x=x_0} \cdots \left. \frac{\partial f(x)}{\partial x_2} \right|_{x=x_0} \right]$$

$$\Rightarrow \nabla_x f(x_0)(x - x_0) = \sum_{i=1}^{n} \left. \frac{\partial f(x)}{\partial x_i} \right|_{x=x_0} (x_i - x_{0,i}) \quad (C.2)$$

defined at point x_0, $\nabla_{xx}^2 f(x_0)$ is the Hessian matrix

$$\nabla_{xx}^2 f(x_0) = \begin{bmatrix} \left. \frac{\partial^2 f(x)}{\partial x_1 \partial x_1} \right|_{x=x_0} & \cdots & \left. \frac{\partial^2 f(x)}{\partial x_1 \partial x_n} \right|_{x=x_0} \\ \vdots & \ddots & \vdots \\ \left. \frac{\partial^2 f(x)}{\partial x_n \partial x_1} \right|_{x=x_0} & \cdots & \left. \frac{\partial^2 f(x)}{\partial x_n \partial x_n} \right|_{x=x_0} \end{bmatrix}$$

$$\Rightarrow (x - x_0)' \nabla_{xx}^2 f(x_0)(x - x_0)$$

$$= \sum_{j=1}^{n} \sum_{i=1}^{n} \left. \frac{\partial^2 f(x)}{\partial x_i \partial x_j} \right|_{x=x_0} (x_i - x_{0,i})(x_j - x_{0,j}) \quad (C.3)$$

at point x_0, and $\varepsilon(x - x_0)$ is the error of approximation (which we typically assumed is bounded). Given this expansion, x_0 defines a maximum if

$$f(x) - f(x_0) = \nabla_x f(x_0)(x - x_0) + \frac{1}{2}(x - x_0)' \nabla_{xx}^2 f(x_0)(x - x_0) < 0$$

$$(C.4)$$

for all x close to x_0. Restricting the focus to functions whose second derivatives are continuous close to x_0, Eq. (C.4) implies two sufficient conditions. First, the vector of first derivatives must vanish

$$\|\nabla_x f(x_0)\|_p \to 0, \quad \|z\|_p \equiv \left| \sum_{i=1}^{n} z_i^p \right|^{\frac{1}{p}}. \quad (C.5)$$

Second, the Hessian matrix is negative definite, or

$$x' \nabla^2_{xx} f(x_0) x < 0 \quad \text{for all } x. \tag{C.6}$$

Newton–Raphson is simply a procedure that efficiently finds the zeros of the gradient vector (a vector valued function) defined by Eq. (C.4). Specifically,

$$\nabla_x f(x) = 0 \Rightarrow \nabla_x f(x_0) - \nabla^2_{xx} f(x_0)(x - x_0) = 0 \tag{C.7}$$

by taking a first-order Taylor series approximation of the gradient vector (as in the Gauss–Sidel procedure discussed in the text). Solving for x based on x_0, we have

$$x = x_0 - [\nabla^2_{xx} f(x_0)]^{-1} \nabla_x f(x_0). \tag{C.8}$$

Thus, by starting from an arbitrary initial point, the analyst can solve for the first-order sufficient conditions by iterating on Eq. (C.8).

Appendix D

An Axiomatic Development
of Expected Utility

This appendix presents the proof in von Neumann and Morgenstern. It attempts to make the proof more accessible than its original terse presentation. However, to facilitate comparisons, the von Neumann and Morgenstern numbering of results have been maintained.

As a starting point, consider the most basic of von Neumann and Morgenstern's conjectures

A:A If $u \prec v$ then $\alpha < \beta$ implies $(1 - \alpha)u + \alpha v \prec (1 - \beta)u + \beta v$.

The direction of the assertion is that if $u \prec v$ and $\alpha < \beta$, then the preference ordering must follow. To demonstrate this, start with axiom 3:B:a: given that $\{0 \leq \alpha \leq 1\}$

$$u \prec v \Rightarrow u \prec \alpha u + (1 - \alpha)v$$
$$\Rightarrow u \prec (1 - \beta)u + \beta v. \tag{D.1}$$

Intuitively, this axiom states that if u is the inferior bundle, then any bundle constructed with any combination of v must be preferred to u. Axiom 3:B:b reverses this axiom by saying that if u is the preferred bundle then it must also be preferred to a bundle containing any amount of v

$$u \succ v \Rightarrow u \succ \alpha u + (1 - \alpha)v. \tag{D.2}$$

Next from Eq. (D.1), replace α with β and replace the first u in the right-hand side with Eq. (D.2) yields

$$(1-\beta)u + \beta v \succ \gamma((1-\beta)u + \beta v) + (1-\gamma)v. \tag{D.3}$$

Next, reverse the order of the combining in the first set of parentheses in Eq. (D.3) yielding

$$(1-\beta)u + \beta v \succ \gamma(\beta v + (1-\beta)u) + (1-\gamma)v. \tag{D.4}$$

Applying the combining axiom 3:C:b to the right-hand side of Eq. (3.A.4) yields

$$(1-\beta)u + \beta v \succ \gamma\beta v + \gamma u - \gamma\beta u + u - \gamma u$$
$$\succ (1-\gamma\beta)u + \gamma\beta v. \tag{D.5}$$

Letting $\alpha = \gamma\beta$ with $1 < \gamma < 1$ completes the proof. Specifically, if $\alpha < \beta$ which is implied by $1 < \gamma < 1$, then the left-hand side of Eq. (D.5) is preferred to the right-hand side.

A:B   Given two fixed points u_0 and v_0 with $u_0 \prec v_0$ consider the mapping

$$\alpha \to w \approx (1-\alpha)u_0 + \alpha v_0.$$

This mapping is a one-to-one and monotone mapping of the interval $u_0 \prec w \prec v_0$.

The mapping results follow the axioms $u \prec v \Rightarrow u \prec \alpha u + (1-\alpha)v$ such that $0 < \alpha < 1$ and $u \succ v \Rightarrow u \succ \alpha u + (1-\alpha)v$. Given this mapping there exists a single number α such that for any number less than α, $w \succ \alpha u_0 + (1-\alpha)v_0$ and for any number greater than α, $w \prec \alpha u_0 + (1-\alpha)v_0$.

A:C   The mapping A:B actually maps the α of $0 < \alpha < 1$ on all w of $u_0 \prec w \prec v_0$.

· The proof is by contradiction. If for some w_0 in the region $u_0 \prec w_0 \prec v_0$ where omitted then for all $\alpha \in (0, 1)$, $(1-\alpha)u_0 + \alpha v_0 \not\approx w_0$ (or $(1-\alpha)u_0 + \alpha v_0 \gtrless w_0$). The region can then be divided into two classes:

Class I: $(1-\alpha)u_0 + \alpha v_0 \prec w_0$

Class II: $(1 - \alpha)u_0 + \alpha v_0 \succ w_0$

Class I is that set of αs that yield a bundle which is *not* preferred to w_0. Class II is that set of αs where the bundle constructed *is* preferred to w_0.

(1) Class I is not empty: 3:B:c $u \prec w \prec v$ implies the existence of $\alpha u + (1 - \alpha)v \prec w$.
(2) Class II is not empty: 3:B:d $u \prec w \prec v$ implies the existence of $\alpha u + (1 - \alpha)v \succ w$.
(3) If α is in Class I and β is in Class II, then $\alpha < \beta$. The classes are disjoint $\alpha \neq \beta$. The alternative would be $\alpha > \beta$.

What is the contradiction? If $\alpha \in$ I and $\beta \in$ II then $\alpha < \beta$. The point of the proof is to demonstrate that $\alpha \not> \beta$. If $\alpha \Rightarrow (1-\alpha)u_0 + \alpha v_0 \approx w_0$ and $\beta \Rightarrow (1 - \beta)u_0 + \beta v_0 \approx w_0^*$ then $\alpha = \beta$ implies $w_0 \approx w_0^*$. Substituting $\alpha = \beta\gamma$ with $0 < \gamma < 1$ yields $w_0 \approx (1 - \beta\gamma)u_0 + \beta\gamma v_0$. If $\gamma = 1$, $w_0 \approx w_0^*$. However, $\gamma < 1$ implies $w_0 \prec w_0^*$. Each counter proof then involves $(1 - \alpha_0)u_0 + \alpha_0 v_0 \prec w_0$ in Class I and $(1 - \alpha_0)u_0 + \alpha_0 v_0 \succ w_0$ is in Class II.

There must exist α_0 with $0 < \alpha_0 < 1$ which separates the classes. Thus, α_0 will be such that for $\alpha < \alpha_0$ the resulting bundle is in Class I and if $\alpha > \alpha_0$ then the resulting set is in Class II.

First, consider α_0 in Class I. Begin by generating the counter point. Specifically, start by trying to generate a new point such that $\alpha \geq \alpha_0$, but $w \prec w_0$. In this case $(1 - \alpha_0)u_0 + \alpha_0 v_0 \prec w_0$ by substitution (i.e., the new bundle w create by α_0 is on the left-hand side) and noting that $w_0 \prec v_0$. Using 3:B:e $u \prec w \prec v \Rightarrow \alpha u + (1 - \alpha)v \prec w$ for some α

$$\gamma((1 - \alpha_0)u_0 + \alpha_0 v_0) + (1 - \gamma)v_0 \prec w_0 \qquad (D.6)$$

assuming $0 < \delta < 1$. Therefore by the combining axiom

$$\gamma((1 - \alpha_0)u_0 + \alpha_0 v_0) + (1 - \gamma)v_0$$
$$\Rightarrow \gamma u_0 - \gamma\alpha_0 u_0 + \gamma\alpha_0 v_0 + v_0 - \gamma v_0$$
$$\Rightarrow \gamma(1 - \alpha_0)u_0 + (1 - \gamma(1 - \alpha_0))v_0. \qquad (D.7)$$

Hence $\alpha = 1 - \gamma(1 - \alpha_0)$ belongs to I. Thus, $\alpha > 1 - (1 - \alpha_0) = \alpha_0$ from the conjecture, but $\alpha \leq \alpha_0$ since $0 < \delta < 1$.

Second, consider α_0 in Class II. Like the scenario above, begin by generating the counter point that $\alpha \leq \alpha_0$, but $w > w_0$. Then $(1 - \alpha_0)u_0 + \alpha_0 v_0 > w_0$ (again substituting a linear combination of α_0 for the bundle w) and $u_0 \prec w_0$. Applying 3:B:d $u > w > v \Rightarrow \alpha u + (1 - \alpha)v > w$ for some α with $0 < \gamma < 1$

$$\gamma((1 - \alpha_0)u_0 + \alpha_0 v_0) + (1 - \gamma)u_0 > w_0. \tag{D.8}$$

Again, combine apply the combining axiom

$$\gamma((1 - \alpha_0)u_0 + \alpha_0 v_0) + (1 - \gamma)u_0$$
$$\Rightarrow \cancel{\gamma u_0} - \gamma\alpha_0 u_0 + \gamma\alpha_0 v_0 + u_0 - \cancel{\gamma u_0}$$
$$\Rightarrow (1 - \gamma\alpha_0)u_0 + \gamma\alpha_0 v_0. \tag{D.9}$$

Note that as $\gamma \to 1$ this expression becomes $w \approx w_0$. However, note that the convergence is from above. Specifically, if $\gamma < 1$ there is more weight on the $(1 - \gamma\alpha_0)$ term ($-\gamma\alpha_0$ becomes less negative). Letting $\gamma\alpha_0 = \alpha$

$$(1 - \alpha)u_0 + \alpha v_0 > w_0. \tag{D.10}$$

Thus, $\alpha > \alpha_0$ which contradicts the original conjecture that $\alpha \leq \alpha_0$.

Thus, A:B and A:C demonstrate $u_0 \prec w \prec v_0$ implies a utility function defined on the numerical interval $[0, 1] \Rightarrow U : (w) \to [0, 1]$. There are some limitations. First, the numerical representation is for a utility interval $u_0 \prec w \prec v_0$, not for all w simultaneously. Second, the numerical representation of A:B and A:C has not been correlated with our requirement 3:1:a and 3:1:b

$$3\!:\!1\!:\!a \;\; u > v \to V(u) > V(v) \tag{D.11}$$

$$3\!:\!1\!:\!b \;\; V(\alpha u + (1 - \alpha)v) = \alpha V(u) + (1 - \alpha)V(v). \tag{D.12}$$

Based on our discussion above, 3:1:a is clearly satisfied. 3:1:b has to be established.

A:D   Let u_0, v_0 such that $u_0 \prec v_0$. For all w in the interval, $u_0 \prec w \prec v_0$ define the numerical function $f(w) = f_{u_0,v_0}(w)$ (f defined on the interval $[u_0, v_0]$) as

(i) $f(u_0) = 0$

(ii) $f(v_0) = 1$

(iii) $f(w)$ for $w \not\approx u_0, v_0$ as α in $0 < \alpha < 1$. Corresponding to the α that divides the set of bundles into Classes I and II.

A:E   This mapping $(w \to f(w))$ has the following properties

(i′) It is monotone,

(ii′) For $0 < \beta < 1$ and $w \not\approx u_0, f((1 - \beta)u_0 + \beta w) = \beta f(w)$, and

(iii′) For $0 < \beta < 1$ and $w \not\approx v_0, f((1 - \beta)u_0 + \beta w) = 1 - \beta + \beta f(w)$.

A:F   A mapping of all w with $u_0 \precsim w \precsim v_0$ on any set of numbers that possesses the properties (i), (ii), and either (ii′) or (iii′) is identical with the mapping in A:D.

A:D is a definition, it must be proven A:E and A:F. Starting with A:E:i′ $w \to f(w)$ is monotone. For $u_0 \prec w \prec v_0$, the mapping is monotone by A:B

$$\alpha \to w \approx (1 - \alpha)u_0 + \alpha v_0 \tag{D.13}$$

is a one-to-one and monotone mapping. Thus, all w of this interval are mapped onto numbers between 0 and 1 ($0 < \alpha < 1$). Hence, the results are monotonic for $u_0 \prec w \prec v_0$.

A:E:ii′   For $w \approx v_0, f((1-\beta)u_0 + \beta v_0) = \beta$ (i.e., letting $f(u_0) = 0$ and $f(v_0) = 1$) this coincides with A:B with β replacing α. For $w \not\approx v_0 \Rightarrow u_0 \prec w \prec v_0, f(w) = \alpha$ by A:B

$$w \approx (1 - \alpha)u_0 + \alpha v_0. \tag{D.14}$$

By 3:C:b $\alpha(\beta u + (1 - \beta)v) + (1 - \alpha)v \approx \gamma u + (1 - \gamma)v \ni: \gamma = \alpha\beta$, and 3:C:a $\alpha u + (1 - \alpha)v \approx (1 - \alpha)v + \alpha u$

$$(1 - \beta)u_0 + \beta w \approx (1 - \beta)u_0 + \beta((1 - \alpha)u_0 + \alpha v_0)$$
$$\approx (1 - \beta\alpha)u_0 + \beta\alpha v_0. \tag{D.15}$$

This is the combining that has been used over and over again. Therefore by A:B $\{\alpha \to w \approx (1 - \alpha)u_0 + \alpha v_0\}$

$$f((1 - \beta)u_0 + \beta v_0) = \alpha\beta = \beta\{\text{as } \alpha \to 1\}. \tag{D.16}$$

A:E:iii′   For $w \approx u_0, f((1 - \beta)u_0 + \beta v_0) = 1 - \beta$ from A:B with $\alpha = 1 - \beta$ and using 3:C:c {i.e., reversing the order in the function}. Think about this. Let w be the lower bound of the interval. If w is the lower bound of the interval $f(w) = 1 - \beta$. For $w \not\approx u_0$ and $u_0 \prec w \prec v_0$ then $f(w) = \alpha$ by A:B ($w \approx (1 - \alpha)u_0 + \alpha v_0$). Using 3:C:b

$$(1 - \beta)v_0 + \beta w \approx (1 - \beta)v_0 + \beta((1 - \alpha)u_0 + \alpha v_0) \tag{D.17}$$

{i.e., since $w \approx (1 - \alpha)u_0 + \alpha v_0$ replace w}. Thus,

$$(1 - \beta)v_0 + \beta w \approx v_0 - \beta v_0 + \beta u_0 - \alpha\beta u_0 + \alpha\beta v_0$$
$$\approx \beta(1 - \alpha)u_0 + (1 - \beta(1 - \alpha))v_0. \tag{D.18}$$

Hence by A:B

$$f((1 - \beta)v_0 + \beta w) = 1 - \beta(1 - \alpha) = 1 - \beta + \beta f(w). \tag{D.19}$$

To reach this conclusion remember that by A:B $\alpha \to w \approx (1 - \alpha)u_0 + \alpha v_0$. Thus

$$w \approx (1 - \beta)v_0 + \beta w \to \alpha \text{ such that}$$
$$(1 - \alpha)(1 - \beta)v_0 + (1 - \alpha)w \approx w^*$$
$$\therefore f((1 - \beta)v_0 + \beta w) = 1 - \beta(1 - \alpha) \text{ \{defines } \alpha\}$$
$$= 1 - \beta + \beta\alpha. \tag{D.20}$$

Thus, by this definition $\alpha \to f(w) \Rightarrow f((1 - \beta)v_0 + \beta w) = 1 - \beta + \beta f(w)$.

To prove A:F, it must be demonstrated that a mapping of all w on the set of numbers possesses the properties:

A:D:i $f(u_0) = 0$,
A:D:ii $f(v_0) = 1$,
A:E:ii′ $f((1 - \beta)v_0 + \beta u_0) = \beta f(w)$, and
A:E:iii′ $f((1 - \beta)v_0 + \beta u_0) = 1 - \beta + \beta f(w)$.

A:1 Consider a mapping $w \to f_1(w)$ with (i), (ii), and either (ii') or (iii').

A:2 Defining the original mapping as $w \to f(w)$.

If this mapping (the original mapping in A:1) is one-to-one, then it can be inverted.

A:3 $\alpha \to \psi(\alpha)$.

Combining A:1 with A:3 (i.e., the inverse mapping of our original definition in A:C) yields A:4

A:4 $\alpha \to f_1(\psi(\alpha)) = \phi(\alpha)$.

The first part of the expression $\alpha \to f_1(w)$ is by A:1. Next, substitute in the inverse mapping set based on $w \to \alpha = f(w) \Leftarrow \alpha = \psi(\alpha)$. Since A:1 and A:2 fulfill (i) and (ii)

A:5 $\phi(0) = 0$ and $\phi(1) = 1$.

Since both A:1 and A:2 fulfill (ii') and (iii')

A:6 $\phi(\beta\alpha) = \beta\phi(\alpha)$ {by ii'}

or

A:7 $\phi(1 - \beta + \beta\alpha) = 1 - \beta + \beta\phi(\alpha)$ {by iii'}

Letting $\alpha \to 1, \{\phi(\alpha\beta) = \beta\phi(\alpha), \phi(\alpha) \to 1\}$ yields

A:8 $\phi(\beta) = \beta$.

Also by letting $\alpha \to 0, \{\phi(1 - \beta + \alpha\beta) = 1 - \beta + \beta\phi(\alpha), \phi(\alpha) \to 0\}$ yields $\phi(\beta) = 1 - \beta$. Considering the definition of $\phi(\alpha)$ {in A:3 and A:4} and the general validity of A:8 which was to be proved.

A:G Let u_0, v_0 be fixed as above and let α_0 and β_0 be fixed with $\alpha_0 < \beta_0$. For all w in the interval $u_0 \precsim w \precsim v_0$ define the numerical function $g(w) = g_{u_0,v_0}^{\alpha_0,\beta_0}(w)$ as follows

$$g(w) = (\beta_0 - \alpha_0)f(w) + \alpha_0$$

($f(w) = f_{u_0,v_0}(w)$ according to A:D). Such that

(i) $g(u_0) = \alpha_0$, and
(ii) $g(v_0) = \beta_0$.

A:H   This mapping $w \to g(w)$ has the properties

(i′) It is monotonic,

(ii′) For $0 < \beta < 1$ and $w \not\succ u_0$ then $g((1 - \beta)u_0 + \beta w) = (1 - \beta)\alpha_0 + \beta g(w)$, and (iii′)

(iii′) For $0 < \beta < 1$ and $w \not\succ v_0$ then $g((1 - \beta)u_0 + \beta w) = (1 - \beta)\beta_0 + \beta g(w)$.

A:I   A mapping of all w with $u_0 \precsim w \precsim v_0$ on any set of numbers which possesses the properties (i), (ii), and either (ii′) or (iii′) is identical with the mapping of A:G.

The proofs of these function is based on

$$g_1(w) = (\beta_0 - \alpha_0)f_1(w) + \alpha_0 \tag{D.21}$$

or

$$f_1(w) = \frac{g(w) - \alpha_0}{\beta_0 - \alpha_0}. \tag{D.22}$$

Since the properties of $f_1(w)$ have been established in A:D-A:F, A:G-A:I follow A:D-A:F.

A:J   Assuming (i), (ii) in A:G the equation $g((1 - \beta)u + \beta v) = (1 - \beta)g(u) + \beta g(v)$ (with $u_0 \precsim u \prec v \precsim v_0$) with $u \approx u_0$ and $v \not\succ u_0$ is equivalent to (ii′) in A:I and with $u \not\succ v_0$ and $v \approx v_0$ is equivalent to (iii′) in A:I.

Appendix E

A GAMS Program to Select Optimal Portfolios

```
sets months Months or observations of returns/
Jan, Feb, Mar, Apr, May,
            Jun, Jul, Aug, Sep, Oct, Nov, Dec/
    stocks Names of stocks/ADM, 3M, HP, JD/;

table data(months,stocks) Monthly returns for
each stock

            ADM        3M        HP         JD

Jan  -0.0282    0.0101    0.0029  -0.0796
Feb  -0.0905    0.0119   -0.0896  -0.0194
Mar  -0.0092    0.0372   -0.0139  -0.0459
Apr   0.0259   -0.0307    0.0482   0.1215
May   0.0857    0.0086    0.1963  -0.0082
Jun   0.0752    0.0198    0.0964   0.0515
Jul   0.0210    0.0870   -0.0061   0.1112
Aug   0.0601    0.0209   -0.0586   0.1128
Sep  -0.0548   -0.0304   -0.0246  -0.0527
Oct   0.0946    0.1419    0.1524   0.1371
Nov   0.0000    0.0063   -0.0255   0.0101
Dec   0.0651    0.0758    0.0603   0.0660

variables
    avg(stocks)      Average return for each stock
    z(stocks)        Holding of each stock
```

```
    portret(months)  Portfolio return for each month
    risk             Objective Function;

equations
    means(stocks)    Calculate the average return
                     for each

stock
    rets(months)     Portfolio return equations
    portbal          Portfolio Balance Constraint
    target           Rate of return target
    variance         Variance;

means(stocks)..
        sum(months,data(months,stocks)/12) =e=
avg(stocks);
portbal..   sum(stocks,z(stocks)) =e= 1.0;
target..    sum(stocks,avg(stocks)*z(stocks))
            =g=0.03; variance..

    sum((stocks,months),(z(stocks)*(data(months,
stocks)- avg(stocks)))
                *(z(stocks)*(data(months,stocks)-
avg(stocks))))
*1/12 =e= risk;
rets(months)..
        sum(stocks,z(stocks)*data(months,stocks))
=e= portret(months);

model prob1 using/all/;
solve prob1 using nlp minimizing risk;

display avg.l;
display risk.l;
display portret.l;
```

Appendix F

R Program to Derive Optimum Portfolio with and without a Risk-Free Asset

```
indta <- read.csv("Chapter 5 Table 5-1.csv")
indta <- as.matrix(indta)
oo     <- c(1,1,1,1,1,1,1,1,1,1,1,1)
# A vector of 12 ones to compute mean
o      <- c(1,1,1,1)
# The vector of 4 ones to be used in
the portfolio computations
mu     <- 0.04
# Set the target rate of return
rf     <- 0.02
# Set the risk-free rate of return
xmean <- t(indta)%*%oo/12
# Compute the mean of returns
xvar  <- t(indta)%*%indta/12-xmean%*%t(xmean)
# Compute the maximum likelihood
variance

print("Mean Rate of Return")
print(xmean)
print("Variance of Rates of Return")
print(xvar)

a      <- t(o)%*%solve(xvar)%*%o
# Compute A coefficient in Equation 5.33
on Page 116
```

```
b       <- t(xmean)%*%solve(xvar)%*%o
# Compute B coefficient in Equation 5.31
on Page 116

c       <- t(xmean)%*%solve(xvar)%*%xmean
# Compute C coefficient in Equation 5.31
on Page 116

print("a coefficient")
print(a)
print("b coefficient")
print(b)
print("c coefficient")
print(c)

# Compute Optimum Portfolio without Risk-Free
Asset

lstar <- (c-mu*b)/(a*c-b*b)
# Compute lambda based on target rate of
return in Equation
# 5.33 on Page 116
print("lambda star")
print(lstar)
gstar <- (mu-lstar*b)/c
# Compute gamma based on portfolio
constraint in Equation 5.31
#    on Page 116
print("gamma star")
print(gstar)
wstar <-
solve(xvar)%*%o%*%lstar+solve(xvar)%*%
xmean%*%gstar
# Compute optimum portfolio using
Equation 5.32 on page 116
print("w star")
print(wstar)
```

```
varport <- t(wstar)%*%xvar%*%wstar
# Compute variance of optimal portfolio
print("optimal portfolio variance")
print(varport)


# Compute Optimum with a Risk-Free Asset


gstar <- (mu-rf)/(c-2*rf*b+rf*rf*a)
# Compute gamma using Equation 5.41
on Page 120
wriskf <- solve(xvar)%*%(xmean-rf*o)%*%
((mu-rf)/(c-2*rf*b+rf*rf*a))
# Compute Optimal Portfolio with Risk-free
asset using
# Equation 5.42 on Page 120
print("optimal portfolio with risk-free asset")
print(wriskf)
varrf <- t(wriskf)%*%xvar%*%wriskf
# Compute the variance of the portfolio
with the risk-free
#    asset
print("risk free portfolio risk")
print(varrf)
```

Appendix G

Program to Compute the Efficient Frontier with and without a Risk-Free Asset

```
indta <- read.csv("Chapter 5 Table 5-1.csv")
indta <- as.matrix(indta)
oo     <- c(1,1,1,1,1,1,1,1,1,1,1,1)
o      <- c(1,1,1,1)
mu     <- 0.05        # Set the target rate of
return
rf     <- 0.01        # Set the risk-free rate of
return
xmean <- t(indta)%*%o/12
xmean <- t(indta)%*%oo/12
xvar   <- t(indta)%*%indta/12-xmean%*%
t(xmean)

a      <- t(o)%*%solve(xvar)%*%o
b      <- t(xmean)%*%solve(xvar)%*%o
c      <- t(xmean)%*%solve(xvar)%*%xmean

for(i in 1:15) {
   xmu    <- i/100
   lstar <- (c-xmu*b)/(a*c-b*b)
   gstar <- (xmu-lstar*b)/c
   wstar <-
solve(xvar)%*%o%*%lstar+solve(xvar)%*%xmean%*%gstar
    varport <- t(wstar)%*%xvar%*%wstar
    if (i <= 1) mm<-xmu else mm<-rbind(mm,xmu)
    if (i <= 1) ll<-lstar else ll<-rbind(ll,lstar)
```

```
    if (i <= 1) gg<-gstar else gg<-rbind(gg,gstar)
    if (i <= 1) ww<-t(wstar) else
       ww<-rbind(ww,t(wstar))
    if (i <= 1) vv<-varport else vv<-rbind(vv,varport)
    }

print(cbind(mm,vv,ll,gg,ww))


for(i in 1:15) {
    xmu      <- i/100
    griskf <- (xmu-rf)/(c-2*rf*b+rf*rf*a)
    wriskf <- solve(xvar)%*%(xmean-rf*o)%*%
                  ((xmu-rf)/(c-2*rf*b+rf*rf*a))
    vriskf <- t(wriskf)%*%xvar%*%wriskf
    z0       <- 1-t(o)%*%wriskf
    if (i <= 1) mm<-xmu else mm<-rbind(mm,xmu)
    if (i <= 1) gg<-griskf else gg<-rbind(gg,griskf)
    if (i <= 1) ww<-t(wriskf) else
       ww<-rbind(ww,t(wriskf))
    if (i <= 1) vv<-vriskf else vv<-rbind(vv,vriskf)
    if (i <= 1) zz<-z0 else zz<-rbind(zz,z0)
    }

print(cbind(mm,vv,gg,ww,zz))
```

GAMS Program for the Portfolio Problem

```
sets icrops crops/1*4/
        consts constraints /income, portf/;

alias(icrops,jcrops);

table var(icrops,jcrops) variance covariance matrix

        1       2       3       4
1 924.41 458.52 202.22 135.22
2 458.52 761.29 452.99   72.55
3 202.22 452.99 490.11 109.09
4 135.22   72.55 109.09 284.17
table line(icrops,const) constraints

   income    portf
1  8.119      1.0
2 11.366      1.0
3  6.298      1.0
4  8.014      1.0
parameter b(const)
      /income 7.00,
       portf 1.00/;
variables  x(icrops)   level of each crop
           rhs(consts) level of each right hand side
           risk;
positive variables x;
```

```
equations   vari    variance
            lconst  linear constraints;
vari..

   sum((icrops,jcrops),var(icrops,jcrops)*x(icrops)
   *x(jcrops)) =e= risk;
lconst(consts)..
   sum(icrops,line(icrops,consts)*x(icrops)) =e=
   b(consts);
model risk1 using /all/;
solve risk1 using nlp minimizing risk;
```

Bibliography

Allias, P.M. (1953). Le Comportement de L'Homme Rationnel Devant Le Risque: Critique des Postulats et Axiomes de L'Ecole Americane. *Econometrica* 21(4)(October): 503–546.

Anderson, J.R., J.L. Dillon and J.B. Hardaker (1977). *Agricultural Decision Analysis*, Ames, Iowa: Iowa State University Press.

Barry, P.J. (1980). Capital asset pricing and farm real estate. *American Journal of Agricultural Economics* 62(3)(August): 549–553.

Black, F. (1972). Capital market equilibrium with restricted borrowing. *Journal of Business* 45(3)(July): 444–455.

Black, F. and M. Scholes (1973). The pricing of options and corporate liabilities. *Journal of Political Economy* 81(3)(May–June): 637–654.

Boussard, J.-M. and M. Petit (1967). Representation of farmers' behavior under uncertainty with a focus-loss constraint. *Journal of Farm Economics* 49(4)(November): 869–880.

Burbidge, J.B., L. Magee and A.L. Robb (1988). Alternative transformations to handle extreme values of the dependent variable. *Journal of the American Statistical Association* 83(401): 123–127.

Chambers, R.G. (1988). *Applied Production Analysis: A Dual Approach*, New York: Cambridge University Press.

Chambers, R.G. and J. Quiggin (2000). *Uncertainty, Production, Choice, and Agency: The State-Contingent Approach*, New York: Cambridge University Press.

Charnes, A. and W.W. Cooper (1959). Chance-constrained programming. *Management Science* 6(October): 78–79.

Chavas, J.-P. and R. Pope (1984). Information: Its measurement and valuation. *American Journal of Agricultural Economics* 66(5)(Decebmer): 705–710.

Cocks, K.D. (1968). Discrete stochastic programming. *Management Science* 15(1)(September): 72–79.

Collender, R.N. and D. Zilberman (1985). Land allocation under uncertainty for alternative specifications of return distributions. *American Journal of Agricultural Economics* 67(4)(November): 779–786.

Collins, R.A. (1985). Expected utility, debt-equity structure, and risk balancing. *American Journal of Agricultural Economics* 67(3)(August): 627–629.

Debreu, G. (1986). Theoretic models: Mathematical form and economic content. *Econometrica* 54(6)(November): 1259–1270.

Diamond, P.A. and J.E. Stiglitz (1974). Increases in risk and in risk aversion. *Journal of Economic Theory* 8(3)(July): 337–360.

Dillon, J.L. and P.L. Scandizzo (1978). Risk attitudes of subsistence farmers in northeast Brazil: A sampling approach. *American Journal of Agricultural Economics* 60(3)(August): 425–435.

Dixit, A.K. and R.S. Pindyck (1994). *Investment under Uncertainity*, Princeton, New Jersey: Princeton University Press.

Fama, E. and J. MacBeth (1973). Risk, return, and equilibrium: Empirical tests. *Journal of Political Economy* 81(3)(May–June): 607–636.

Featherstone, A.M. and C.B. Moss (1990). Quantifying gaines to diversification using certainty equivalence in a mean variance model: An application to florida citrus. *Southern Journal of Agricultural Economics* 22(3)(December): 191–198.

Featherstone, A.M., C.B. Moss, T.G. Baker and P.V. Preckel (1988). The theoretical effects of farm policies on optimal leverage and the probability of equity loss. *American Journal of Agricultural Economics* 70(3)(August): 572–579.

Freund, R.F. (1956). The introduction of risk into a programming model. *Econometrica* 24(3)(July): 253–263.

Gloy, B.A. and T.G. Baker (2002). The importance of financial leverage and risk aversion in risk-management strategy selection. *American Journal of Agricultural Economics* 84(4)(November): 1130–1143.

Hardar, J. and W.R. Russell (1969). Rules for ordering uncertain prospects. *American Economic Review* 59(1): 25–34.

Hazell, P.B.R. (1971). A linear alternative to quadratic and semivariance programming for farm planning under uncertainty. *American Journal of Agricultural Economics* 53(1)(February): 53–62.

Huang, C.-F. and R.H. Litzenberger (1988). *Foundations for Financial Economics*, New York: North-Holland.

Ingersoll, J. (1987). *Theory of Financial Decision Making*, Totowa, New Jersey: Rowman & Littlefield Publishers, Inc.

Just, R.E. and R.D. Pope (1978). Stochastic specification of production functions and economic implications. *Journal of Econometrics* 7(1)(February): 67–86.

Just, R.E. and Q. Weninger (1999). Are crop yields normally distributed? *American Journal of Agricultural Economics* 81(2)(May): 287–304.

Kamien, M. and N.L. Schwartz (1991). *Dynamic Optimization: The Calculus of Variations and Optimal Control in Economics and Management*, New York: Elsevier.

Koenker, R. and G. Bassett (1978). Regression quantiles. *Econometrica* 46(1): 33–50.

Kroll, Y., H. Levy and H.M. Markowitz (1984). Mean–variance versus direct utility. *Journal of Finance* 39(1)(March): 47–61.

Lintner, J. (1965a). Security prices, risk, and maximal gain from diversification. *Journal of Finance* 20(December): 587–615.

Lintner, J. (1965b). The valuation of risk assets and the selection of risky investments in stock portfolios and capital budgets. *Review of Economics and Statistics* 47(February): 13–37.

Malliaris, A.G. and W.A. Brock (1982). *Stochastic Methods in Economics and Finance*, New York: North-Holland.

Machina, M.J. (1987). Choice under uncertainty: Problems solved and unsolved. *Journal of Economic Perspectives* 1(1)(Summer): 121–154.

Markowitz, H.M. (1959). *Portfolio Selection: Efficient Diversification of Investments*, New Haven, CT: Yale University Press.

May, K.O. (1954). Intransitivity, utility, and the aggregation of preference patterns. *Econometrica* 22(1)(January): 1–13.

Meyer, J. (1975). Increasing risk. *Journal of Economic Theory* 11(August): 119–132.

Meyer, J. (1977a). Choice among distributions. *Journal of Economic Theory* 14(April): 326–336.

Meyer, J. (1977b). Second degree stochastic dominance with respect to a function. *International Economic Review* 18(2)(June): 477–487.

Meyer, J. (1987). Two-moment decision models and expected utility maximization. *American Economic Review* 77(3)(June): 421–430.

Mirowski, P. (1991). The when, the how, and the why of mathematical expression in the history of economic analysis. *Journal of Economic Perspectives* 5(1)(Winter): 145–157.

Moss, C.B., A.P. Pagano and W.G. Boggess (1994). Ex ante modeling of the effect of irreversibility and uncertainty on citrus investments. *Risk Modeling in Agriculture: Retrospective and Prospective Program. Proceedings for the annual meetings of the Technical Committee of S-232*, Department of Economics, Iowa State University, Ames, August.

Moss, C.B., A.M. Featherstone and T.G. Baker (1987). Agricultural assets in an efficient multiperiod investment portfolio. *Agricultural Finance Review* 49: 82–94.

Moss, C.B., S.A. Ford and M. Castejon (1991). Effect of debt position on the choice of marketing strategies for Florida orange growers: A risk efficiency approach. *Southern Journal of Agricultural Economics* 23(2)(December): 103–111.

Moss, C.B. and G. Livanis (2009). Implementation of stochastic dominance: A nonparametric kernel approach. *Applied Economics Letters* 16(15)(October): 1517–1522.

Moss, C.B., R.N. Weldon and R.P. Muraro (1991). The impact of risk on the discount rate for different citrus varieties. *Agribusiness: An International Journal* 7(4)(July): 327–338.

Moss, C.B. and J.S. Shonkwiler (1993). Estimating yield distributions with a stochastic trend and nonnormal errors. *American Journal of Agricultural Economics* 75(4)(May): 1056–1062.

Mossin, J. (1966). Equilibrium in a capital asset market. *Econometrica* 34(October): 768–783.

Nelson, C.H. and P.V. Preckel (1989). The conditional beta distribution as a stochastic production function. *American Journal of Agricultural Economics* 71(2)(May): 370–378.

Paris, Q. and C.D. Easter (1985). A programming model with stochastic technology and prices: The case of Australian agriculture. *American Journal of Agricultural Economics* 67(1)(February): 120–129.

Pratt, J.W. (1964). Risk aversion in the small and in the large. *Econometrica* 32(1/2)(January–April): 122–136.

Preckel, P.V., A.M. Featherstone and T.G. Baker (1987). Interpreting dual variables with nonmonetary objectives. *American Journal of Agricultural Economics* 69(4)(November): 849–851.

Ramirez, O.A. (1992). A new specification for economic modeling of non-normal dependent variables. Staff Paper 91–10, Food and Resource Economics Department, University of Florida, May.

Ramirez, O.A., C.B. Moss and W.G. Boggess (1994). Estimation and use of the inverse hyperbolic sine transformation to model nonnormal correlated random variables. *Journal of Applied Statistics* 21(4): 289–304.

Raskin, R. and M.J. Cochran (1986). Interpretations and transformations of scale for the Pratt–Arrow absolute risk aversion coefficient: Implications for generalized stochastic dominance. *Western Journal of Agricultural Economics* 11(2)(December): 204–210.

Ross, S. (1976). The arbitrage theory of capital asset pricing. *Journal of Economic Theory* 13(3)(December): 341–360.

Rothschild, M. and J.E. Stiglitz (1970). Increasing risk I, a definition. *Journal of Economic Theory* 2(3)(September): 225–243.

Schmitz, A. (1995). Boom/bust cycles and Ricardian rent. *American Journal of Agricultural Economics* 77(5)(December): 1110–1125.

Schmitz, A., C.B. Moss, T.G. Schmitz and H. Furtan (2009). *Agricultural Policy, Rent Seeking, and Global Interdependence*, Toronto: University of Toronto Press (forthcoming).

Shackle, G.L.S. (1949). *Expectations in Economics*, Cambridge, England: Cambridge University Press.

Shackle, G.L.S. (1961). *Decision, Order, and Time in Human Affairs*, Cambridge, England: Cambridge University Press.

Shannon, C.E. and W. Weaver (1963). *The Mathematical Theory of Communicatin*, Chicago: University of Illinois Press.

Sharpe, W.F. (1964). Capital asset prices: A theory of market equilibrium under conditions of risk. *Journal of Finance* 19(September): 425–442.

Theil, H. (1967). *Economics and Information Theory*, Amsterdam: North-Holland Publishing Co.

Turvey, C.G., C.L. Escalante and W. Nganje (2005). Developments in portfolio management and risk programming techniques for agriculture. *Agricultural Finance Review* 65(2): 219–245.

Tversky, A. (1969). Intransitivity of preferences. *Psychological Review* 76(January): 31–48.

von Neumann, J. and O. Morgenstern (1953). *Theory of Games and Economic Behavior*, Princeton, NJ: Princeton University Press.

Index

American option, 212
arbitrage portfolio, 196, 197
Arbitrage Pricing Model, 7
Arbitrage Pricing Theory (APT), 183, 196
Arrow Pratt absolute risk aversion, 110, 111
Arrow-Pratt risk aversion, 156
Arrow-Pratt risk aversion coefficient, 89–91

basis, 210, 211, 226
Bayes Theorem, 19
Bayes' equation, 174
Bernoulli distribution, 84
beta, 193, 194
beta distribution, 23

call, 211–216, 219
Capital Asset Pricing Model (CAPM), 7, 109, 129, 183, 190, 191, 194–196
capital market line (CML), 187, 190
cash price, 210, 211
certainty equivalent, 7, 58, 63–66, 84, 85, 88, 96–98, 128, 145, 146, 195, 196
certainty equivalent approach, 195
choice criteria, 4
Cobb-Douglas, 5, 59, 60, 69, 146

conditional distribution, 22
contingent assets, 211
continuous random variable, 12
controllable inputs, 228, 231
correlation coefficient, 21
covariances, 21, 33, 186
cumulative distribution, 151–154, 157–160, 170–172

deterministic, 4, 11
discrete, 11–13
discrete random variables, 11
distribution function, 21–23, 35–37

efficiency criteria, 7, 150–152
efficient portfolios, 106, 110
element, 13
entropy, 177–180, 182
European option, 212, 214
event, 4, 5, 11–13
expected utility, 6, 124, 126, 128, 139, 141, 146
expected utility hypothesis, 65–69, 150
expected value, 28–31
expected value-variance, 73, 124, 129, 131, 134, 146, 150, 185, 196
expected value-variance formulation, 102, 110–112, 122

expected value-variance frontier, 106, 108, 109, 113, 119, 120
expenditure function, 58, 60, 62, 63

factor model, 196, 197
first-degree stochastic dominance (FSD), 151, 152, 167
Focus-Loss, 136–138
frequency approach, 12, 13
futures markets, 209

gamma distribution, 22, 39, 47, 50
Gauss–Markov theorem, 46
Gauss–Sidel, 49

hedging, 209
Hicksian demand, 61, 62
homoscedasticity, 46

idiosyncratic risk, 197
increasing risk, 154–157, 160
independence, 11, 19–22, 31, 41
independently and identically distributed, 41, 46
indirect utility function, 58–60, 62, 63

marginal distribution, 18, 22
Marshallian demand, 60
martingale, 52, 55
mean preserving spread, 157
mode, 25
moment generating function, 33, 41
MOTAD, 131, 135, 137, 138, 141, 143

negative exponential, 102, 110
negative exponential utility, 89, 91, 128
normal distribution, 20, 21, 23, 24, 33, 34, 38, 39, 41, 43–46

objective function, 4
observation, 11, 41
option pricing approach, 7
outcome, 11–13, 16, 17, 39

portfolio choice, 125, 126
posterior probability, 174, 178
power utility function, 63, 65, 69, 84, 85, 89, 90
prior probability, 173, 174, 178
probability, 4, 12–19, 21, 36, 38, 52
probability density function, 214, 224
put, 211–213

random, 10–13
random variable, 11–13, 15–18, 20, 23, 24, 28–30, 33, 36, 52
real options, 209
risk adjusted discount rate (RADR), 194, 196
risk aversion, 7, 83–85, 87–99, 175–177
risk efficiency, 150, 172
risk efficiency criteria, 7
risk neutral, 83, 85, 176, 177, 212
risk preferring, 83
risk premium, 58, 64, 84, 85, 88

sample space, 11, 15–17, 21, 36, 52
second-degree stochastic dominance (SSD), 153, 167, 169, 172
security market line (SML), 191, 194
set, 11–16, 20, 37, 52, 53, 55
statistic, 28
stochastic, 11, 55
stochastic dominance with respect to a function, 154
stochastic production function, 227–229, 231
stock option, 209
strike price, 8

time to maturity, 214
transformation functions, 23
treatment, 11
triangular probability density, 25, 26

uncontrollable factors, 228
uniform distribution, 21, 22, 29, 32, 39

quadratic utility function, 70